*design of regional accounts*

# *Design of REGIONAL ACCOUNTS*

*Papers presented at the Conference on Regional Accounts, 1960*

*Sponsored by the* COMMITTEE ON REGIONAL ACCOUNTS

*Edited by Werner Hochwald*

PUBLISHED FOR RESOURCES FOR THE FUTURE, INC.
BY THE JOHNS HOPKINS PRESS, BALTIMORE

Manufactured in the United States of America

Standard Book Number 8018-0277-6

*Originally published,* 1961
Second printing, 1969

## *Committee on Regional Accounts*

(as of December 1960)

Resources for the Future is a nonprofit corporation for research and education in the development, conservation, and use of natural resources. It was established in 1952 with the co-operation of The Ford Foundation and its activities have been financed by grants from the Foundation. Part of the work of Resources for the Future is carried out by its resident staff, part supported by grants to universities and other nonprofit organizations. Unless otherwise stated, interpretations and conclusions in RFF publications are those of the authors; the organization takes responsibility for the selection of significant subjects for study, the competence of the researchers, and their freedom of inquiry.

This book is an outcome of one of RFF's regional studies, which were then directed by Harvey S. Perloff, and have been directed since 1968 by Lowdon Wingo. Werner Hochwald is Professor of Economics at Washington University, St. Louis.

RFF *staff editors*, Henry Jarrett, Vera W. Dodds, Nora E. Roots, Sheila M. Ekers.

# PREFACE

THE IDEA THAT regional accounts could become useful systems of information flow for public and private decision-making has been evolving over a long period of time. Several years ago, Dr. Harvey S. Perloff, director of regional studies for Resources for the Future, Inc., initiated an organized effort to bring this idea to fruition. Resources for the Future invited me to spend a year with its regional studies group to study central problems in regional accounts and approaches to their development. The objective was to see whether enough conceptual progress in the field of economics had been made to hold out hope for the development of a system of regional accounts; also availability of data needed to be explored. Subsequently Resources for the Future made a grant to Washington University to enable the work to be continued.

In May 1958 a score of scholars met to discuss the prospects for a more extensive and continuous effort. This resulted in formation of the Committee on Regional Accounts, which has been active ever since.

The Committee on Regional Accounts decided to sponsor a number of seminars and conferences that would permit scholars to exchange ideas and help advance the field. The papers presented in this volume are the product of the first conference, which had been preceded by a three-day seminar, held in December 1959, where the various authors presented and discussed early versions of their papers. The Conference on Regional Accounts took place on September 7 and 8, 1960, on the campus of Washington University, St. Louis.

The field of regional accounts draws heavily on developments in the fields of national accounts as well as regional analysis. It attempts to identify significant questions that need to be answered on the subnational level, and develops a system of accounts which will help present this information in an orderly manner.

The speed with which imaginative research on urban and regional problems will move forward will depend, in large part, on the early development of a powerful conceptual framework and data to implement it. Availability of appropriate systems for data presentation, together with the data themselves, can greatly affect the quality of decisions which will be made in years to come with regard to urban and regional problems.

From the papers presented in this volume it appears clear that the information requirements will include demographic, political, and economic data, and the description of the region's main assets. A well-articulated system of information flow should prove of help to local public officials in weighing the effects of alternative policy proposals on the region's health and well-being. In addition, regional analysis and accounts, prepared in a consistent manner in many regions, should make much valuable information available to private decision-makers throughout the country and thus improve the general level of private decision-making and the efficiency of the national economy. Similarly, a workable system of regional accounts could help provide an informational base on which federal and state policy-makers could make better decisions consistent with both local and national interests.

It is the hope of the Committee on Regional Accounts that the papers presented in this volume will stimulate many others to enter the field. If scholars take some of these papers as a point of departure for further fruitful work, the preparation of this volume has been well worth while.

The Committee would like to express its deep appreciation to Dr. Reuben G. Gustavson, who was president of Resources for the Future, Inc., when this project was initiated, and to Dr. Joseph L. Fisher, its present president. Both encouraged our efforts and provided helpful guidance. We also are especially grateful to Dr. Werner Hochwald, who so effectively and resourcefully served as editor of this volume.

Werner Z. Hirsch

# CONTENTS

# INTRODUCTION

ACCOUNTS TO RECORD information flows are as old as written history. In times past, when change was slow, their main function was to enter debits and credits for the purpose of preserving authentic evidence of legal relations among transactors; this basic function is still reflected in the modern language of accounting. Temple accounts to record contributions of the faithful are found in the oldest documents of literate man. Business accounts to protect creditors and absentee stockholders have given rise to the profession of certified public accountants. Government accounts to restrict the arbitrary use of public funds reflect the long struggle for parliamentary democracy which has shaped the modern budgetary process.

As change has been accelerating and economic growth has become a major social objective, accounts have added to their purely protective function the task of helping to bring about these very changes. The history of private capitalism is intimately interwoven with the development of double-entry bookkeeping which relates profit and loss to changes in assets and debt. Business accounting has thus been going through a continuous process of adaptation to become a more useful tool of management for the projection of profits as they determine and are determined by the discounted value of investment in capital stock. This adaptation has been facilitated for the individual firm by the clear definition of its legal relations with the outside world, by the market evaluation of its business assets, and by the certain criteria of immediate profit goals.

None of these conditions are met for the nation as a whole, and national economic accounts therefore show a much more complex history.[1] The first adaptation of business accounts to national policy is found in the balance of trade when mercantilists identified the nation with the royal household whose goal was to maximize treasure. More recently, with the revived interest in monetary policy and liquidity preferences, a more sophisticated "balance of trade" among domestic sectors of the national economy has been presented in flow of funds accounts,[2] while periodic shifts in the structure of claims are shown by a national balance sheet.[3] A more continuous history can be traced for the long attempts, starting with Davenant's *Use of Political Arithmetic* (1698), to summarize the confusing variety of economic data in a single net index of national welfare, such as national income or national wealth. Reflecting the Keynesian concern with multiplier effects of strategic macroeconomic variables, such as consumption, business investment, and government expenditures, these single indices have now been expanded into a consistent set of national income and prod-

[1] Design and use of national economic accounts are discussed in the *Studies in Income and Wealth* published by the Conference on Research in Income and Wealth, now 24 volumes (Princeton: Princeton University Press). See also the *Income and Wealth* series published by the International Association for Research in Income and Wealth, now 8 volumes (Chicago: Quadrangle Books). The International Association sponsors in addition to the above series a *Bibliography on Income and Wealth,* now 7 volumes (Chicago: Quadrangle Books). For a history of national economic accounts see Paul Studenski, *The Income of Nations* (New York: New York University Press, 1958). Review, appraisal, and recommendations are contained in *The National Economic Accounts of the United States,* Hearings before the Subcommittee on Economic Statistics of the Joint Economic Committee, Eighty-fifth Congress, October 29 and 30, 1957. Standard texts are Richard Ruggles and Nancy D. Ruggles, *National Income Accounts and Income Analysis* (New York: McGraw-Hill, 1956); John P. Powelson, *National Income and Flow-of-Funds Analysis* (New York: McGraw-Hill, 1960).

[2] Morris A. Copeland, *A Study of Moneyflows in the United States* (New York: National Bureau of Economic Research, 1952). Official estimates are published in *Flow of Funds in the United States, 1939-1953* (Washington: Board of Governors of the Federal Reserve System, 1955), and current issues of the *Federal Reserve Bulletin.* For international data see *International Financial Statistics,* issued monthly by the International Monetary Fund.

[3] Raymond W. Goldsmith, *A Study of Saving in the United States* (Princeton: Princeton University Press, 1955). Also by the same author, *Financial Intermediaries* (Princeton: Princeton University Press, 1958).

uct accounts.[4] The further need to trace gross production flows through their successive stages in the national economy has been recognized since Quesnay's *Tableau Economique* (1758). Recent attempts to implement a Walrasian system of general equilibrium analysis are presented in Leontief's input-output table[5] to show the interindustry relations impact of a stipulated change in the net final demand for the nation's output of goods and services.

Regions as well as industries may be viewed as sector aggregates in a national accounting system. The resources of the nation are of uneven spatial mobility, and national economic analysis has always been concerned with the spatial advantages of individual regions within the nation, a concern stimulated by high transportation costs. Many of the data compiled for the nation since the days of political arithmetic, such as early population and wealth estimates, have been classified by location. Petty's *The Political Anatomy of Ireland* (1672) can be viewed as a study in regional comparative advantage concerned with explaining the flow of mobile factors, such as liquid funds and skilled workers, to join geographically fixed resources, such as land. National-regional-urban structural relationships as they affect the location of industry have been studied in a more formal setting since von Thünen's *Der Isolierte Staat* (1826).[6]

[4] Simon Kuznets, *National Income: A Summary of Findings* (New York: National Bureau of Economic Research, 1946). For an official presentation of U. S. data see *U. S. Income and Output* (Washingon: U. S. Department of Commerce, 1958). Current data are published in the *Survey of Current Business.* For international data see the *Yearbook of National Accounts Statistics* issued by United Nations.

[5] W. W. Leontief, *The Structure of the American Economy, 1919-1939* (New York: Oxford University Press, 1951). For an exposition of the official U. S. data see W. Duane Evans and Marvin Hoffenberg, "The Interindustry Relations Study for 1947," *The Review of Economics and Statistics,* May 1952. A standard text is Hollis B. Chenery and Paul G. Clark, *Interindustry Economics* (New York: Wiley, 1959).

[6] For a survey of location theory see Walter Isard, *Location and Space-Economy* (New York and Cambridge: Technology Press and Wiley, 1956). For a historical study of regional differentials in United States economic growth see Harvey S. Perloff, Edgar S. Dunn, Jr., Eric E. Lampard, and Richard F. Muth, *Regions, Resources, and Economic Growth* (Baltimore: Johns Hopkins Press, 1960). See also *The New York Metropolitan Region Study* directed by Raymond Vernon and edited by Max Hall (Cambridge: Harvard University Press, 1960).

Two problems stand out in the study of a region. First, the very definition of the region offers many ambiguities. Its size may vary from the small neighborhood within a larger metropolitan area to the huge subregion within a continent. As the very purpose of regional analysis is the study of resource immobilities, the geographic boundaries of transportation flows suggest themselves for defining a region, such as a commuting or trading area. Yet as the extent of spatial immobilities varies with the resources and time periods chosen, no single definition appears satisfactory, and the "region" itself may change in the very process of regional development. The most obvious compromise is a compilation of data by small administrative units, such as counties, to serve as "building blocks" for various types of regions. Even a small administrative unit may be too large for certain purposes, however, and Harvey Perloff emphasizes in his paper on Relative Regional Economic Growth the importance of studying localization patterns within the region in addition to the forces determining industrial location among different regions.

The definition of a region remains ambivalent for still another reason. As the appropriate size of a region varies with the particular resources on which it is focused, there are many overlapping circles of regional interests and identification. The small neighborhood area is focused on daily household needs within walking distance while the large trading region is determined by geographic and technological considerations of efficient transportation flows. These different circles of regional interest identification are paralleled by a confusing multiplicity of overlapping institutional jurisdictions, ranging all the way from the small school district with less than fifty pupils and the neighborhood store organized as a single proprietorship, via corporate businesses of regional size and the traditional state governments, to the national corporation and federal agencies without regional identification. This rich variety of spatial institutions appears essential to a free pluralistic society though it may also encourage fragmentation, isolationism, and parochialism.[7] Where the region is well integrated into the nation, the individual citizen participates in every one of these multitu-

[7] For an interesting discussion of these problems see Robert C. Wood, *Metropolis Against Itself* (New York: Committee for Economic Development, 1959).

dinous institutional layers of local and national identification. An important aspect of this institutional multiplicity is the effective separation of property interests from local physical immobilities through the operations of a national money and capital market. A region within the same geographic boundaries may be defined either in terms of the property and other institutional interests of its residents, or in terms of specific physical limitations to resource mobility, with quite different implications. The region within a well-integrated national economy thus lacks the precise meaning which it is possible to stipulate for the individual business firm or the nation as a whole.

This ambiguity in the definition of a region leads to the second problem of regional analysis. The smaller the region, and the more integrated the national economy, the wider it is open to outside influences which will determine its economic activities. In this respect the region is like a business firm continuously adjusting itself to external changes beyond its control, rather than like a nation driven by its own internal forces of economic development. Regional data[8] therefore receive much of their analytic usefulness from comparisons with the data of other regions, and regional analysis is essentially the study of comparative regional advantage to estimate the future role of the region in the national economy. Methods of regional analysis are thus concerned with the establishment of a logical framework for comparative data to measure and project economic change; these methods range from a simple juxtaposition of socioeconomic profiles taken from various regions to more elaborate statistical constructs of time and cross-section series, industrial complex analysis, and spatial interaction models.[9]

In this volume, Harvey Perloff surveys the different approaches to regional analysis most of which emphasize the external forces as they shape the growth and development of individual regions, with the analytical focus "from the outside in." Edgar Hoover and Benjamin Chinitz call this approach "horizontal" in their compara-

[8] One hundred and thirty-three different types of current local data have been compiled in the *County and City Data Book*, U. S. Bureau of the Census, 1957.

[9] For a comprehensive survey see Walter Isard, *Methods of Regional Analysis* (New York and Cambridge: Technology Press and Wiley, 1960).

tive structural analysis of the Pittsburgh Metropolitan Region. Similarly, Louis Delwart and Sidney Sonenblum discuss regional analysis in the context of national projections. In all these cases the process of regional development is visualized as a continuous effort to bring about favorable shifts in the regional share of the national economy. Strategic changes are expected to occur in the regional "export" industries serving the national market from which base the activities of purely "local" industries are derived.

Regional accounts, just as national income accounts, have had their beginning in the search for a single yardstick to summarize the wide variety of local economic data. Regional personal income has served as such a measure to express in a single term the many facets of local development. State and county incomes have been estimated in response to the needs of business and government for a single index of local purchasing power and tax capacity.[10] Werner Hirsch, in the lead article of this volume, explores the possibilities of expanding personal income into a more comprehensive index of health and well-being which would use a multiple set of criteria to indicate the goals and accomplishments of the region. In addition to past income, such criteria include expected future income,[11] employment stability, net social benefits, and less tangible amenities of life.

[10] The official state income estimates since 1929 have been analyzed by Charles F. Schwartz and Robert E. Graham, Jr., in *Personal Income by States* (Washington: Department of Commerce, 1956). Current annual estimates are published each year in the August issue of the *Survey of Current Business,* monthly unofficial estimates in *Business Week.*

John Littlepage Lancaster gives a careful bibliography of early county income estimates in *County Income Estimates for Seven Southeastern States* (Charlottesville: University of Virginia, 1952). A technical supplement to this report has been prepared by Lewis C. Copeland, *Methods for Estimating Income Payments in Counties* (Charlottesville: University of Virginia, 1952).

Problems of regional income estimation are discussed in *Regional Income,* Vol. 21 of the Studies in Income and Wealth (Princeton: Princeton University Press, 1957). A more detailed bibliography on regional accounts is contained in *Local Impact of Foreign Trade* (Washington: National Planning Association, 1960).

[11] Burton A. Weisbrod, in a forthcoming article on "An Expected-Income Measure of Economic Welfare," has suggested ways to measure the value of human capital as an index of regional growth.

On the national level, income estimates have been extended into a closed system of income and product accounts to show how expenditures of past income determine product flows. On the regional level, such an approach appears less fruitful as the view "from the outside in" suggests that strategic production decisions are made in response to expected "foreign" market demand rather than local expenditures of past income. Regional accounts therefore are "open" to trace the impact of external forces on local production flows; their design depends on the choice of exogenous and endogenous sectors to trace the relations between the region and the "rest of the world." Werner Hirsch in his paper develops a "regional interaction model" whose main "inputs" are local public policies and external forces while its "outputs" are the dimensions of the region's health and well-being.

To understand the problems of implementing such a model it will be helpful to recall once more the basic elements of any accounting system. Accounts are defined by their transactors and the transactions they record. In business accounting, transactors are always legal entities whose transactions, in a double-entry system, are related to changes in their balance sheets. In economic accounting, transactors may be either sector aggregates or legal entities, as in all moneyflows and many income accounts, or they may be physical constructs, such as the "industry" in input-output accounts. Transactions may again be related to asset changes, as is done for moneyflows accounts, or they may be recorded without explicit reference to changes in stocks, as is done for income and product accounts.

Any regional accounting system, to form an internally consistent framework, must therefore carefully define its components and their aggregation into endogenous or exogenous sectors. As regions are usually defined in terms of physical immobilities rather than institutional properties, "real" income and product or input-output accounts can be adapted more readily to regional analysis than "monetary" flow of funds accounts. The measurement of "real" flows in interregional trade is most likely to serve the primary purpose of regional accounts, i.e., to estimate comparative regional advantage. Where resource control rather than its physical location forms the focus of interest, however, accounts defined in terms of the regional claims structure would appear more appropriate.[12]

Richard and Nancy Ruggles show in their paper how the various types of national economic accounts can be broken down into regional accounting systems. Charles Leven constructs an imaginative set of regional income and product accounts to record local trade relations with the "rest of the world" in a "from-to" matrix. Louis Delwart and Sidney Sonenblum use regional accounts to trace the impact of external forces on regional employment and thus to project regional activities in the context of national demand projections. Edward Berman sketches the outlines of a more specialized regional interindustry model to show the impact of local government policies on capital formation and output. Data gaps and the resultant problems of measurement, as they affect the accounting design, are stressed by most of the contributors. Britton Harris, in his discussion of the Hoover-Chinitz Pittsburgh study, pleads for a much-needed national census of transportation to provide more adequate "from-to" interregional flow data.

As regional analysis is concerned with understanding the region's relative economic advantage in supplying the national demand for goods and services, a statement of regional resources must be linked with the flows of regional product. Such linkage of flow and stock accounts presupposes some valuation of regional assets which is most readily available for monetary claims as they are recorded in flow of funds accounts. Yet monetary claims are by definition mobile and therefore of less interest for a statement of spatial immobilities. The region's relative advantage is determined by its fixed resources which may attract mobile factors from other parts of the country. George Borts and Jerome Stein, in their paper on investment return, use marginal productivity theory to evaluate regional assets and thus to predict relative growth differences among regions as well as industrial growth patterns within the region.

Regional accounts and their potential contributions must be

[12] John Stuart Mill, in his *Principles of Political Economy* (1848), draws a clear distinction between resource location and property interests: "If Manchester, instead of being where it is, were on a rock in the North Sea (its present industry nevertheless continuing), it would still be but a town of England, not a country trading with England, it would be merely as now, a place where England finds it convenient to carry on her cotton manufactures." (Book 3, chap. 25, p. 685 in the Ashley edition.)

weighed against the costs and insights offered by other methods of regional analysis. Many of the authors and discussants therefore also comment on their limitations. Edgar Hoover and Benjamin Chinitz point to the high cost of implementing an accounting system complete enough to permit meaningful internal consistency checks. Similarly, Jesse Burkhead in his discussion of the Perloff paper indicates the cost advantages of conceptually more rudimentary data which are easily communicable to decision-makers. On the other hand, no accounting system, however complete, will reveal all the behavior determinants of regional transactors. Guy Orcutt suggests a microanalytic model, solved by simulation methods, to study more intensively the aspects of micro-components which are either difficult to aggregate or which lose much of their significance when aggregated in accounting sectors. Similarly, Roland Artle in his discussion of the Hirsch paper pleads for more understanding of individual decision units and their interaction. Yet, as James Petersen points out in his discussion of the Orcutt paper, the number of variables to be accommodated even in the most ambitious microeconomic model is limited, and the need therefore remains to find categories appropriate for the aggregation of microrelations.

Regional accounts are neither supposed to answer all questions nor are they required for all occasions. Like business and national economic accounts, they are most useful when they are comprehensive and comparable over time. Regardless of data limitations, regional accounts offer a flexible framework for economic analysis, a framework which permits piecemeal implementation as the research resources at hand permit and the policy issues justify. Regional accounts are therefore also most helpful to indicate priorities for further research efforts. Even where the accounting model remains nonoperational because of persistent data gaps, regional accounts encourage and facilitate a more explicit and precise statement of the multitudinous assumptions which inevitably underlie any regional projection. No one would expect business accounting to substitute for the art and skills of business management. Yet the language of accounting is helpful to communicate and express business policies and business projections.

To put the matter still differently, business and national economic accounts have in common a framework to describe relations among transactors who are defined individually for business accounts and grouped into sector aggregates for economic accounts. The "region" provides a new set of criteria for sector aggregation, based either on physical location or ownership controls. Variations in spatial mobility, and overlapping institutional jurisdictions, make these aggregation criteria for an "open" regional economy less definite than they are for a "closed" national economy. Yet, though less definite than national transactor groupings, regional aggregation criteria offer a new dimension to describe the complex interactions of the national economy. The contributors to this volume have taken up the challenge of designing such a more generalized system of economic accounting.

Werner Hochwald

# 1

# A GENERAL STRUCTURE FOR REGIONAL ECONOMIC ANALYSIS

*Werner Z. Hirsch**

*Washington University*

MOST AMERICAN economists point with pride to our system of national accounts as one of their greatest achievements. Since 1932, when the U. S. Senate requested the Department of Commerce to prepare estimates of national income jointly with the National Bureau of Economic Research, a powerful analytical framework has been developed which is backed by the systematic collection of pertinent information, and which serves as a basis for economic decisions by the Council of Economic Advisors, the Federal Reserve Board, and Congress, and also by private policy-makers. To be sure, there is not always complete unanimity either in the interpretation of the data or in the rating of priorities. But in the formulation of fiscal and monetary policies these basic differences are often brought into the open and debated in the light of up-to-date data considered in the context of an internally consistent framework. Data and framework are today so universally understood and so closely followed by private groups that it is unthinkable for the federal government to delay action unduly or act in a grossly irrational manner.

* The author is indebted to Harvey S. Perloff, Harold J. Barnett, Roland Artle, and Martin Beckmann, for constructive suggestions and criticism on earlier drafts of this paper.

On the other hand, in the eyes of most observers, decisions with regard to regional economies—whether cities, counties, metropolitan areas, or states—have been generally poor, inadequate, or ineffectual. Complaints most commonly voiced against state and local officials are that they are poorly informed and that their decisions are more often influenced by what is politically palatable than by rational considerations of what is best for community and country. Complaints against federal legislators and officials are that many of their actions are carried out without a clear appreciation of their regional implications. The fact is that on the subnational (i.e., regional) level, a framework for economic policy-making does not exist, and the data are all too often scanty, faulty, or altogether absent. In many communities scattered throughout this country, efforts have been initiated by public and private groups to collect data and establish some kind of background for the policies of local governments and businesses and for the guidance of voters. But these efforts have not produced a framework and system of data collection for the field of regional and local economies comparable to that which has proven so valuable on the national level.

Thus, this paper rests on the premise that the construction of an integrated conceptual framework for subnational economies and the development of a system of data collection for its implementation are not only desirable but feasible. Whether these efforts are termed "regional accounts" or whether a less forbidding term can be invented is of minor consequence. In any case, the nature of the account system must be dictated by the problems it is designed to help solve. Since it will be shown below that regional and national problems are different, it should become evident that the emphases of regional and national account systems differ also.[1]

[1] This viewpoint disagrees with that taken by Ezra Solomon and Zarko G. Bilbija in their *Metropolitan Chicago* (Chicago: Chicago Free Press, 1959). They applaud the contributions national accounts have made to improved governmental and business planning on the national level. Then they proceed to attempt a regional breakdown of certain portions of national accounts in the hope of making metropolitan economic analyses. There can be no doubt that gross regional income and product measures and some breakdowns can depict basic characteristics of the Chicago situation and indicate some temporal changes; but in themselves they are of limited use in identifying opportunities for industrial expansion and in estimating the impact of private and public decisions affecting the region.

A region's economic future depends to no small extent upon its human and natural resources, its economic structure, the capacity and vigor of the national economy, government's ability to create a favorable environment, and the actions, traditions, and aspirations of private decision-makers. Regional analysis in general, and regional accounts in particular, must be designed to improve private and public policy-makers' understanding of these forces and their interaction. In the public sector state and local officials make decisions about such things as public services, expenditures and taxes, transportation facilities, and the use of land and of other resources such as water, forests, and parks. In these efforts they can be materially aided by the engineer and physical planner on the one hand, and by the social scientist on the other. Thus, in order to decide on the desirability of specific objectives and a plan to implement them, benefit-cost analyses are needed. However, the social scientist can develop cost estimates only after detailed plans have been worked out. He can make his contribution by actually estimating the probable costs and benefits of the plan—after having identified and provided measurements of the main dimensions of a "region's health and well-being." This information can then be used by the engineer and physical planner to modify the original proposal or to advance alternatives, which can then be subjected again to benefit-cost analyses by the social scientist. In this manner, all three join in producing a number of alternative plans together with their respective benefit and cost estimates.

Decisions by the federal government tend to have local repercussions, some of which may not be anticipated; examples are the Federal Housing and Highway Acts. Other federal decisions might be made with the explicit objective of establishing an environment within which local public and private decision units are more likely to act in a manner consistent with the nation's best interest. Or they might be designed to combat differential changes in regional economic activity, which could become particularly important should partial disarmament become a reality.

There is general interest, too, in the development of projections of regional economic activity. Such projections, among other benefits, can improve the quality of private location and marketing decisions.[2]

To help understand, trace, and possibly measure the numerous complex interrelations between local public and private policies and external forces, and to assess how their impact ultimately affects the region's health and well-being, a "regional interaction model" will be constructed.[3] Among the main 'inputs" of this model are local public policies and external forces (including state and federal policies); its "outputs" are the dimensions of the region's health and well-being, which will be developed below in detail.

This step will be followed by a presentation of instruments that offer hope for the implementation of parts of the regional interaction model. Major emphasis will be on regional flow analysis, since papers by Harvey S. Perloff, George H. Borts and Jerome Stein, and Edward B. Berman will address themselves to aspects of asset or regional relative economic advantage analysis.

Finally, some early empirical results will be examined and some thoughts expressed about particularly challenging and significant steps which are deemed important for the future.

## *I. REGION'S HEALTH AND WELL-BEING*

While economists have long been concerned with welfare, i.e., economic well-being, this paper uses the term, "health and well-

[2] For excellent discussions of many of the issues faced in making regional projections, see Raymond Vernon, *Metropolis 1985* (Cambridge: Harvard University Press, 1961), as well as Jerome L. Stein and Mark B. Schupack, *Forecast of New England's Machinery Industry in 1970* (Boston: Federal Reserve Bank of Boston, 1959).

[3] Roland Artle, in his formal comments on this paper, stated, "I have a bias against general systems, general theories, general structures. It seems what is needed mostly today in economics is not to build upwards, toward general systems and theories, but rather to build downwards, toward some understanding of the individual decision-making units and their actions and interactions."

The author would like to express his agreement with Professor Artle about the importance of gaining a better understanding of decision-making units. But such steps appear to be easier once a framework exists, which helps identify the more significant decision-making questions and places important questions into proper relations. Thus, if there exists a difference in the two positions, it is merely one of strategy.

being." This somewhat broader term includes, for example, amenities of life, usually not considered part of economic well-being.

Perhaps the most commonly used yardsticks for measuring the status of economic well-being of the national economy have been gross national product and national income. But are these two measures appropriate and, particularly, are they useful criteria for the determination of a region's health and well-being? The answer is an emphatic "No!" Instead, it appears promising to use a multiple set of criteria—i.e., the most significant properties by which the status of a region in two different time periods can be compared.

In our philosophy of society, the health and well-being of a region are made up of the health and well-being of its individuals. So, before drawing up a list of major significant dimensions of a region's health and well-being, the main items that impinge upon the utility function of its individuals need to be identified.

Basically, health and well-being stem from the consumption of goods and services (1) produced and bought from the private sector, (2) produced and bought from the public sector, and (3) originating in nature, nonprofit institutions, and the general environment. As an approximate measure of the utility of Class 1, per capita personal real income together with employment stability can be used, while net social benefits can measure the utility of Class 2, and economic growth together with amenities of life represent that of Class 3.

### *Dimensions of Health and Well-being*

In line with these considerations, the following five main dimensions of a region's health and well-being can be identified:

a) per capita personal real income,
b) basic employment stability,
c) net social benefits,
d) economic growth, and
e) amenities of life.

Next, some major characteristics of a region's health and well-being and means to quantify them will be explored in some detail.

*Per Capita Personal Real Income.* Perhaps the single best indication of how well off the average person is at a given point in time is per capita personal income. This measure can be improved by isolating from it price level changes. Thus, by deflating per capita disposable income in current dollars by an index reflecting price level changes—e.g., the Bureau of Labor Statistics' Consumer Price Index—per capita personal real income can be estimated.

There are a number of complicating measurement issues. Perhaps the simplest one is that allowance needs to be made for home-grown food and the services a home-owning family receives from its home without paying rent. A much more difficult matter centers around age distribution and household size.[4] Furthermore, some knowledge about income distribution is desirable. This is hard to get and perhaps even harder to appraise. While increases in per capita personal real income are indubitably welfare gains under existing conditions in the United States, changes in income distribution do not lend themselves to such an unequivocal interpretation.

State personal income data are at present prepared on an annual basis by the U. S. Department of Commerce's Office of Business Economics, and on a monthly basis by *Business Week.* The Office of Business Economics is working on annual personal income estimates for major metropolitan areas.

*Basic Employment Stability.* Quarterly or, preferably, monthly regional employment and unemployment data, if properly interpreted, can eloquently testify to the extent of cyclical stability. This stability, or rather its counterpart, instability, has two dimensions: (1) duration, and (2) extent of departure from some norm.

George H. Borts has developed and applied some interesting statistical techniques of analysis to annual state employment data.[5] Without too much difficulty, they promise to be adaptable to the

[4] For an interesting exposition of these points see Martin David, "Welfare, Income and Budget Needs," *Review of Economics and Statistics,* 41 (November 1959), 393-99.

[5] George H. Borts, "Regional Cycles of Manufacturing Employment in the United States 1914-1953," *Journal of American Statistical Association,* 289 (March 1960), 151-211.

type of problems and data encountered in analyzing regional employment stability.[6]

However, for many purposes it is more important to learn about and reason in terms of a region's basic employment stability. This involves an identification of the characteristics which contribute to a region's being inherently more stable or unstable. Specifically, whether a region's economy is basically more or less stable depends on the importance of the exogenous segment industries, the nature of each of the exogenous segment industries, the relative importance of output going to the exogenous segment, and the magnitude of employment multipliers of exogenous segment industries. Thus, these four elements need to be studied.

[6] On this point, Harold J. Barnett, in his formal discussion of this paper, stated: "With respect to cyclical employment stability as Professor Hirsch develops it, I can offer criticisms. Professor Hirsch means to guide communities assisting economic development through acquisition of new plants. After the fact, all of us would prefer more cyclical stability to less *other things equal.* But this is not the choice. It is before the fact and with other things not equal that industry location encouragement is given or not. In an industry viewed as unstable, like aircraft production, one can get a very stable high-wage plant. Perhaps McDonnell Aircraft is an example. Contrariwise, in an industry viewed as stable, one can get cyclical instability.

"The automobile and heavy machinery midwest have had instability of employment but also high income and rapid growth. From Professor Borts' recent article, I learned that Michigan has been cyclically very unstable in employment, but has very high per capita income and rapid growth. Yet the present employment problem in autos has become more secular than cyclical.

"Some formerly unstable industries, like aircraft and electrical and electronic goods, have had the capacity to stimulate local industries of the 'desirable' type, such as advanced electronics and research and development.

"Of course, I am overstating. But the economic dimension for economic health and well-being that I'd prefer is absence of durable unemployment. The employment conditions adverse to economic well-being in West Virginia, Kentucky, New England, and Detroit are due to structural change."

The first part of the criticism appears to rest on a misunderstanding. This paper does not assume "other things (dimensions of health and well-being) equal." On the contrary, it attempts to focus on the need to have explicit choices between alternative policies made by trading increases in one dimension against decreases in a second.

As to the second part, many economists would agree with Professor Barnett when he says that he prefers "absence of durable unemployment." However, durable or secular unemployment comes under the general heading of "economic growth" (or decline) and deserves to be separated from cyclical unemployment, which is best treated under the heading of "basic employment stability."

Finally, Professor Barnett is right when he calls for more attention to the duration of unemployment as against the amplitude of the cyclical swing.

*Net Social Benefits.* No doubt, per capita personal real income is a useful first approximation of economic well-being. However, people are forced to give up parts of their income in the form of taxes, and in return they receive certain public services. Furthermore, there is no assurance that benefits from services will be commensurate with tax payments; as a matter of fact, it is quite likely that major discrepancies exist.

Net social benefits (or costs) of government services can be defined as the difference between what people should be willing to pay for public goods and services (consistent with benefits received) and what they actually pay. The concern here is with net social benefits accruing to the people of a region.

Before the question of what gives rise to net social benefits can be examined, the equilibrium position must be presented. In the abstract, at any one moment of time there is an equilibrium position at which the region maximizes its net social benefits of public goods

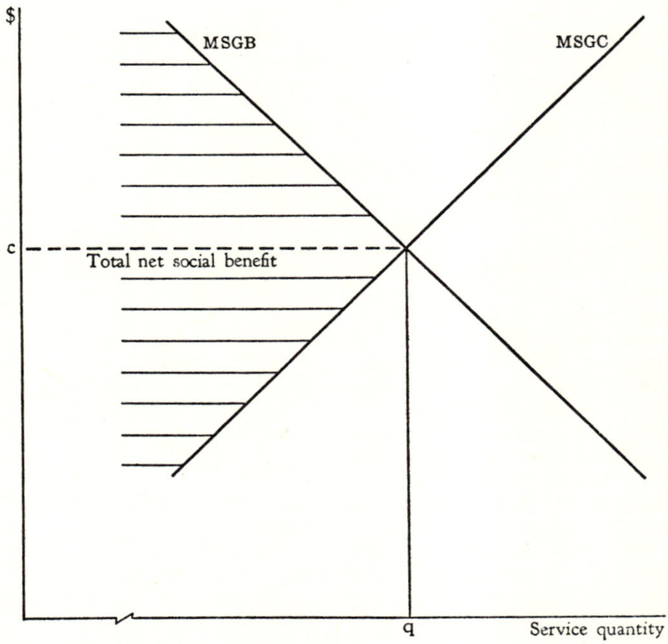


*Figure 1.* Net social benefit.

(or minimizes its net social losses). It is at a point where the region's marginal social government costs and benefits are equal, i.e., where MSGC = MSGB. The net social benefits of the region are represented by the area between MSGC and MSGB, lying to the left of their intersection. (See Figure 1.)

It might be argued that virtually every industry can produce net social benefits. Why then should we single out the public sector? The answer is in terms of some rather unique characteristics of public goods. They tend to have no substitutes and people have no alternative but to buy them.[7]

A region's social benefits may not be maximized, i.e., the public sector may be operating to the left or right of the equilibrium point. Two main reasons might be responsible—imperfect information and implementation difficulties.

Social cost and benefit functions of the public sector are very difficult to estimate. While many knotty conceptual problems remain to be solved, there are even more empirical difficulties. Thus, estimates are likely to be imprecise and biased. Furthermore, there is the temptation to substitute for social costs and benefits more readily available accounting costs and benefits. The interception of the marginal accounting cost and benefit functions is likely to be at a different point than the interception of the functions which would produce an optimum solution.

In the public sector the translation of knowledge into action is especially cumbersome and fraught with delays and imperfections. Distortions tend to occur since decisions about public goods are made within the political arena. Even if the pulling and pushing that typify the political process are based on excellent information and produce a proposal to voters consistent with the equilibrium solution at the time such recommendations are made, often years will pass until implementation is completed. In the meantime, conditions in our dynamic world are likely to have changed.

There are a host of initial stimuli—internal as well as external ones—which can produce changes in these functions and, consequently, change the region's net social benefits from the public

[7] In the extreme, public goods have no substitutes whatever. Good examples are national defense and services of the judiciary.

sector. Methods need to be developed which can help appraise possible effects of initial stimuli on the net social benefit status of a region.

This approach is distinctly different from the usual way of looking at local finance problems. The futility of arguing about a region's objective and image merely in terms of low taxes and high-quality service is all too clear; the two are almost mutually exclusive. Also, there remains the issue: whose taxes should be low and whose services numerous and of high quality?

*Economic Growth.* Economic growth occupies a unique position, for the residents of a region benefit from growth primarily, if not exclusively, to the extent that it improves the environment within which people live and work. In many respects, it resembles an amenity of life.[8] This point can perhaps best be illustrated in the following manner. In the United States we have always had a rapidly growing society. Our institutions, with minor exceptions, have been geared to growth. Growing regions offer superior opportunities to the young and able. Consequently, in the past, those regions which grew at a substantially slower pace than the rest of the nation proved unattractive to many who turned toward greener pastures, usually in rapidly growing regions.

Such a view of regional growth is a major and basic departure from most of the writings on urban and regional economics. Virtually all such writings, explicitly or implicitly, pay almost unlimited tribute to growth. They fail to recognize that, except for its environmental aspects, a region's growth is more a means than an end, and must be judged in terms of the resulting changes that it brings about in the other dimensions of the region's health and well-being. For example, the rapid economic growth of California apparently has had the short-run effect of keeping per capita incomes lower than they otherwise would have been. On the other hand, relatively slow growth or no growth can lead to per capita income increases, at least in the short run. For example, between 1939 and 1954, the states that had absolute decreases in popula-

[8] The author would like to acknowledge his gratitude to Harvey S. Perloff, Tibor Scitovsky, and Martin Beckmann, who in conversation helped shed light on this issue.

tion—such as Oklahoma, Arkansas, and Mississippi—had some of the greatest relative increases in per capita income in the nation for the period.[9] Likewise, rapid growth of industries subject to volatile demand can have an adverse effect on a region's basic employment stability. In those countries in which growth has virtually ceased and whose political and economic institutions have so adjusted themselves, regional economic growth can be almost solely a means to an end.

Appropriate growth measures could be population, labor force, employment, output or value added. With minor exceptions, all these measures are, on the average, closely correlated. The best employment series is produced by the Bureau of Labor Statistics, since 1952 for selected metropolitan areas. The data are derived from a sampling of nonfarm establishments for which payroll data are compiled on the number of employees who received pay for any part of a specified period. Output and "value added" figures for manufacturing become available only every four years from the Census of Manufactures. Reliable population data are produced even less frequently.[10]

*Amenities of Life.* Among the various dimensions of a region's general condition, amenities of life are the most difficult to define and quantify. Yet it is hoped that a number of disciplines jointly can shed light on this complicated issue. For convenience sake, amenities of life, especially in metropolitan regions, might be looked upon as conditions resulting from the services of nature, nonprofit institutions, and the general environment. In a sense they are the grand total of such items, mainly, as quality of nongovernmentally financed educational, recreational, and cultural

[9] Harvey S. Perloff, "Lagging Sectors and Regions of the American Economy," *American Economic Review*, 50 (May 1960).

[10] Growth data, too, must be interpreted with care. For example, employment data are greatly influenced by the region's industrial structure: the garment and leather products industries employ many women who are the family's second breadwinner. At the same time, dominance of the steel industry in a region tends to lead to underemployment of the female labor force, as is well illustrated in Pittsburgh. As a matter of fact, it is interesting to consider whether a region like Pittsburgh would want to attract industries which make fuller use of the female labor force. Or is it perhaps more desirable for women not to be gainfully employed?

facilities; ease of circulation as related to the layout and topography of the metropolitan region; quality of living and working conditions; and general civic spirit.[11] This enumeration claims to be neither all-inclusive nor devoid of overlapping. However, great care is required to distinguish these amenities of life from goods and services produced strictly by the government or private sector.[12]

### *Region's Value of Human Capital*

A few words should be said about an alternative approach. Instead of basing the health and well-being analysis on vector comparisons, the concept of a region's value of human capital has been suggested as an alternative. The net value of human capital has been defined as "the sum of the person's expected future productivity stream (in money terms) net of his consumption, and discounted to the present."[13] Expected future productivity is consid-

[11] Intangible aspects of urban life are well discussed in John E. Burchard, "The Urban Aesthetics," *The Annals of the American Academy of Political and Social Science,* 314 (November 1957), 112-22.

[12] In connection with amenities of life, Harold J. Barnett stated: "The discussion on the fifth dimension, 'amenities of life,' was deficient. Most of the items have already entered economic health and well-being through per capita income. Regarding services of nature and civic spirit, these could usefully undergo economic analysis. For example, California and Michigan both have nature and both Los Angeles and Detroit lack civic spirit, and I think these are correlated—you can't attend a PTA meeting on water skis or in a duck blind. But these features are not ready for a regional interaction model."

The author would want to take issue with the first part of Professor Barnett's comment. The contention that "most items (amenities of life) have already entered (into the computation of) economic health and well-being through per capita income" appears unwarranted in many metropolitan area problems. For example, two heads of households may be working in the same downtown office, holding similar positions, and yet they may live, and often do live, in different municipalities or parts of the metropolitan area, each offering distinctly different amenities of life. The amenities of life associated with their place of residence will hardly affect their income.

Whether amenities of life and civic spirit are inversely related is an interesting question and deserves attention.

[13] Burton A. Weisbrod, *The Value of Human Capital,* paper presented to a meeting of the Southern Economics Association, Jacksonville, Fla., November 20, 1959.

The author would like to express his sincere appreciation to Werner Hochwald, Burton A. Weisbrod, and the other members of the Washington University Seminar on Urban and Regional Economics who are responsible for advancing and clarifying issues related to the concept of the value of human capital.

ered a function of value of output for each sex during each year of age, and the probabilities of surviving each age.

The advantage of this approach would be that the analysis could proceed in terms of a single index number. The disadvantages appear to be threefold. First, a value of human capital index does not incorporate all the important dimensions discussed earlier. Weisbrod has tentatively indicated that the value of human capital reflects current and expected future income, expected future employment level and stability, future life expectancy, and expected duration of work losses due to sickness during a lifetime. The second disadvantage stems from the need to agree on an appropriate interest rate to discount the future. Finally, there is the issue whether a region's health and well-being are more appropriately measured by a net or gross measure reflecting the value of human capital. In the gross measure, consumption would not be deducted from the productivity stream. There apparently is no clear answer to this question. In an affluent society in peacetime, a gross measure might tend to be superior to a net measure.

## *II. REGIONAL INTERACTION MODEL*

Once the main dimensions of a region's health and well-being can be quantified, a powerful model is needed to relate different stimuli–inputs—to these characteristics—outputs. For example, a simulation model should help estimate the specific contributions alternative public policies can be expected to make to each of the dimensions of the region's health and well-being.

A word of caution appears appropriate at this time. It would indeed be unrealistic to hope that some kind of marginal analysis will become applicable—i.e., to define the best possible policy for the region as that which results in equating long-run marginal costs and benefits, where costs and benefits are declines and increases in the region's over-all health and well-being respectively. The problems of measurement and evaluation are too great to make such a formal approach feasible. Instead, perhaps the most promising approach is one of quantifying the impact of alternative policies on the specific dimensions of the region's health and well-being, and letting public opinion or the wisdom of public officials decide which one offers the most desirable outcome under the circumstances.[14]

With these considerations in mind, an attempt will be made to build a model which is applicable to a variety of original stimuli, although the more significant ones are local, state, and federal policies as well as some other external forces. Private sector policies are mainly conceived as adjustments in the light of these original stimuli. The general outlines of the regional interaction model are presented in Figure 2, which might be looked upon as representing snapshots of the model's four successive stages.

### *Stage I*

This stage represents a solution to the system prior to the application of a stimulus. The region's characteristics and their relationships are assumed to be known. It is necessary to describe private and public sector environment (or stock of assets), and interindustry flow relations in period $t$.[15] Furthermore, the major external forces that affect the local economy, i.e., those that originate outside the system, need to be described. The main items under the three headings—i.e., local public sector environment, local private sector environment and external forces—are given in Figure 2. These are the elements that are carefully examined by local and nonlocal private decision units when they search the region's economy for new as well as vanishing business opportunities.

### *Stage II*

The local economy is subjected to an initial stimulus. Mainly, stimuli in the form of local public sector environment (policy)

[14] It appears that the Upper Midwest Economic Study is sympathetic to this view and will attempt to judge the desirability of alternative action programs in just this manner. See James M. Henderson, *The Upper Midwest Economic Study: A Research Prospectus* (Minneapolis: University of Minnesota, 1960), pp. 24-25.

[15] Throughout this paper the analysis of local environmental factors or stock of asset is briefly referred to as asset or relative economic advantage analysis. It should be recognized that the term "asset" here does not have its conventional meaning.

*Figure 2.* REGIONAL INTERACTION MODEL*

| | | Private Sector Environment | Public Sector Environment | External Forces |
|---|---|---|---|---|
| I | Interindustry Flow Relations | a. Manpower<br>per capita income, basic employment stability, employment, etc.<br>b. Capacity<br>c. Production and marketing techniques<br>d. Raw materials & processed goods<br>e. Transportation facilities<br>f. Credit facilities<br>g. Markets | a. Public Services<br>b. Capacity<br>c. Taxation<br>d. Traffic and transit<br>e. Land use planning and regulation<br>f. Planning & regulation of other than land resources | a. Level of economic activity<br>b. Resource picture<br>c. Transportation<br>d. Technology and productivity<br>e. Federal and state legislation<br>f. Local legislation in other regions |
| | Initial Stimulus Occurs in the Form of Local Legislation or External Force Changes and Leads to: | | | |
| II | 1st Change in Interindustry Flow Relations | 1st Change in Private Sector Environment | 1st Change in Public Sector Environment | |
| | Immediate Adjustment by Private Decision Units Has Direct Impact and Leads to: | | | |
| III | 2nd Change in Interindustry Flow Relations | 2nd Change in Private Sector Environment | 2nd Change in Public Sector Environment | |
| | Further Adjustment by Private Decision Units Has Indirect Impact and Leads to: | | | |
| IV | 3rd Change in Interindustry Flow Relations | *New Status of Region's Health and Well-being:*<br>Final Change in Private Sector Environment<br>*Per capita income*<br>*Basic employment stability*<br>*Employment*<br>*Amenities of Life* | Final Change in Public Sector Environment<br>*Net social benefits* | |

* This particular chart assumes a complex initial stimulus constituting and/or causing changes in flow relations, private and public sector environment, and external forces.

changes and external forces are of interest to this analysis. A submodel is needed to permit estimation of the relationships between various stimuli and private and public sector environment and interindustry flow relations in period $t + 1$. While the stimulus can have an impact on many environmental factors (assets), only those need to be included that are likely to be considered by private decision-making units in formulating policies with regard to the region's economy. As a result of the stimulus, and sometimes concomitantly with it, changes in one or more of the important characteristics of the region's environment take place. For example, large-scale changes in land-use regulation can alter the amount of land available for the private sector; this can lead to a change in private sector environment and, simultaneously, it may affect the general government environment as seen by the private decision-maker. Much federal and state legislation has about the same effect where the region is concerned. As another example, local tax increases will lead to a change in public sector environment as well as to a change in interindustry flow relations.[16]

[16] In this connection Roland Artle made the following formal comment: "It seems to me that Werner Hirsch has spent too little time on the relations between the 'inputs' and the 'regional interaction model.' For instance, there is quite a distance between changes in land use regulation or in local taxes and the fact that a new plant will be located in the area concerned. In manipulating the model, Hirsch simply *assumes* that a new plant is located in the area, without going into the connections between this event and the 'inputs.' "

This point is well taken. It might be helpful to take the plant location example and suggest an approach to the problem. In a sense it involves a regional relative advantage analysis, a subject taken up by the following three papers. As is indicated in the discussion of Stage III, plant location decisions in relation to the regional interaction model may be best understood if the analysis proceeds along three lines—a careful study of the special needs of the industry in relation to the assets of the region; a careful study of the decision-making machinery of the industry; and an imaginative study of the value judgments which tend to enter decisions.

Once a given plant is added, the change in regional interindustry flow relations can be approximated by a corresponding adjustment in the input-output table. The changes in the private sector environment are both on the demand and supply side; and it might be useful to follow this dichotomy to study them. A similar approach could be used in relation to changes in the public sector environment.

Admittedly, more work is needed to link inputs with the regional interaction model.

*Stage III*

Private decision-making units face a changed local environment and adjust themselves to it. These immediate adjustments have a direct impact on the region's economy. In this connection it is necessary to describe the considerations which enter into firms' decisions with regard to the region's economy. This is perhaps best done by looking at the problem in the following manner: As a result of a region's changed relative economic advantage, firms that do or could operate in the region face a new demand and supply situation. This new situation must be related to the external force—over-all level of economic activity. In this manner a more or less formal system evolves for estimating the new final demand facing the region's industries.

More specifically, it may be said that in Stage III, industrial, trade and service, and household units examine the new local environment that was created in Stage II. Each unit has specific needs, value judgments, and a particular type of decision-making machinery. In the light of all three, the region's relative economic advantages and the level of national economic activity are reassessed and an adjustment to the new environment takes place. To understand needs, value judgments, and decision machinery, careful industry studies are needed. Policy decisions by industrial and trade and service units lead to output adjustments, while household policies affect labor supply, and with it, availability of skills and wage levels. As industrial, trade and service, and household units make their output and supply adjustments, public and private sector environment changes occur.

While it is important to determine the local economic activities which will be generated directly by the change in local public policy or external forces, it is in practice most difficult, if not impossible, to separate them from those indirectly induced. Many private decision units will attempt to anticipate the direct effects and will adjust themselves to them in advance. In spite of the empirical difficulty of distinguishing between the direct and indirect effects, it is conceptually most helpful to make such distinction.

*Stage IV*

The indirect effects of the public policy change are permitted to work themselves out. Private and public activities, directly generated by the change in local public policy or in external forces, offer numerous new local business opportunities. The mere expectation of these opportunities will tend to generate secondary effects. The extent to which the inputs of the new direct output will be supplied by the local economy will vary greatly. Except for labor inputs, virtually all other inputs may be either all locally produced or all imported. Most likely the outcome will tend to be somewhere between these two extremes.

When the various local private and public decision-making units face the changed local environment of Stage III, each one attempts to adjust itself to these new conditions. While one eye is on the modified local environment, the other remains on the external forces. The various interactions and adjustments bring about final changes in the interindustry flow relation, inputs, outputs, and public and private sector environment, i.e., a new state of the region's health and well-being is reached. At the end of Stage IV a new solution to the system has been attained.

As was done in Stage I, the different building blocks of the public and private sector environment might be quantified in Stage IV. However, for most purposes it will suffice to quantify merely the five main dimensions of a region's health and well-being. Making comparisons of these five dimensions in Stages I and IV, i.e., by using comparative static analysis, affords an appraisal of the impact of the specific local public policy or change in external forces of Stage II on the region's health and well-being.

Actually, the new status of a region's health and well-being can now be called Stage I′ and can be subjected to a further initial stimulus. Again we trace it through Stages II′, III′, and IV′.

Although so far a single stimulus has been considered, in theory there is no reason why it would not be possible to deal simultaneously with a number of different stimuli. For example, technology tends to change continuously, and so does the level of economic activity. In addition, a specific public policy might be introduced. The great difficulty with simultaneously handling a number of stimuli rests with the fact that little is known in a quantitative manner about the nature of the interaction of different stimuli.

The regional interaction model visualizes a comparative static analysis, with the time span of the various stages differing significantly. Stage II is virtually instantaneous. The length of Stage III will be determined by how long it takes decision units to make immediate and direct adjustments to the modified environment. Minor changes in the scale of the unit's operation, for example, should require relatively little time. On the other hand, if a new plant needs to be built, up to a year or two might elapse until the adjustment is completed. Finally, Stage IV will tend to extend over a rather long period before all indirect changes can fully work themselves out.

## *III. INSTRUMENTS OF IMPLEMENTATION*

Following an exposition of the proposed regional interaction model, which admittedly is merely an exploratory conceptual framing of significant issues in the political economics of regions, issues relative to its possible implementation will be taken up. In this connection it is important to state that this paper does not claim to provide an operational model.[17] As a matter of fact, there is no assurance at this time that a unified operational model can be developed and data for its implementation procured. It appears useful, however, to search for operational models which cover portions of the over-all framework. This is the objective of the following pages.

[17] The author is in full agreement with Harold J. Barnett, who in his formal discussion of the paper stated: "The paper can also be appraised as an operational model—and Professor Hirsch unintentionally lays himself open to criticism on this score. On this count, I can only say, quite flatly, that it is not an operational model—mathematically, data-wise, or in economic definition or rigor. The input-output table component of Professor Hirsch's total system is an operational model, as we know from the Leontief innovation and amplifications by Hirsch and others. But this is the only part of the system presented which is. I can amplify this, but it is unnecessary because I do not think that Professor Hirsch really viewed his paper this way."

### *Assets and Flows*

Before such a step is taken, one further look will be given to the regional interaction model, to reflect once more on the interaction between assets and flows. Perhaps one of the more promising approaches is to look upon Stage I of the regional interaction model as representing a solution to the system (prior to the application of a stimulus) in terms of a regional input-output model. The region's interindustry flow relations are represented by the technical coefficient matrix or the inverted matrix; the external forces together with the region's private and public sector environment determine the region's final demand matrix. To put it differently, the general level of economic activity and the relative economic advantage of the region in meeting national (and international) demands for specific goods and services determine its final demand.

While this highly formalized view is not the only possible one, it appears especially useful when, with its help, the other stages of the model are examined. For example, the injection of an initial stimulus at the beginning of Stage II can now be looked upon as mainly producing changes in the region's final demand and technical coefficients. The advantage of looking at the problem in this manner is apparent in its ability to identify important linkage relationships for which submodels must be built and ultimately implemented empirically. The submodel would identify the various possible stimuli and quantify their potential impact on final demand and technical coefficients. In the abstract, once the new final demand and revised technical coefficients have been determined, output and income impact estimates result directly from multiplying final demand by the inverted matrix values. In theory, any number of different initial stimuli can be fed into such a simulation model, and with its aid the prospective impact can be studied in a general way. Useful insight into employment stability and net social benefit impacts can be gained with the aid of additional side calculations.

If the initial stimulus leads to changes in the region's assets and flow picture, which in turn leads to changes in technical coefficients and final demand, the latter two issues deserve further investigation. Much more work is needed to gain a sound understanding of how specific external forces or local government policies modify a region's technical coefficients. Work in this area is likely to be time-consuming and perhaps expensive, but the conceptual problems appear manageable as long as only short-term changes are considered.

The same does not hold for relating an initial stimulus to final demand changes. Here conceptual problems loom large. Because three papers on different aspects of this subject are also presented here, this paper will only touch on the subject of a systematic approach to regional advantage analysis. It merely will apply very broad strokes to reflect on some potentially fruitful approaches. However, let us remember that, in the fullest sense of the terms, "asset" or "regional relative economic advantage analysis," and "goods and service flow analysis" are destined to make important contributions to the further development of regional economic analysis.

### *Regional Relative Economic Advantage Analysis*

Much experience has been accumulated in taking inventory of local stocks of assets, be they in the public or private realm. Likewise, much is known about the environmental requirements of various industries. Still, more careful pinpointing of crucial dimensions and more purposeful analysis within a systematic framework remain to be accomplished.

Several methods appear to offer powerful ingredients for a systematic approach to analyzing a region's assets and relative economic advantage, which in turn can facilitate regional final demand projections. At least two such methods are developed in detail in later papers. Thus, merely preliminary notes will be given here on various methods which in all probability will have to be used in conjunction.

Harvey S. Perloff and associates have developed an approach to regional analysis based on "share analysis," "shift analysis," and "input-output access analysis," which together offer a useful framework for the determination of a region's relative economic advantage.[18] In this connection it is useful to remember that in the regional interaction model, the region's final demand in Stage I is (assumed) known and the main concern centers around estimating changes during the ensuing stages. Thus, with the region's share of national demand in Stage I known, "input-output access" analysis is designed to help project the region's share.

Another method centers around "return on investment," i.e., the income that results from the investment of an additional dollar in new plant and equipment. Presumably the extent of investment undertaken in a given region depends upon the rate of return that private policy-makers expect to earn. Thus, these decision-makers would consider past experience of different industries in different regions and modify it by analyzing and projecting regions' main assets.[19]

Walter Isard and associates have developed a method which they refer to as "industrial complex analysis."[20] It combines an evaluation of interindustry relations, along input-output lines, with an analysis of the factors that lead to regional differences in costs and revenues, along industry comparative-cost lines. This method, which has been applied to Puerto Rico, explicitly recognizes that major industrial development tends to be vertically and circularly

[18] Harvey S. Perloff, Edgar S. Dunn, Jr., Eric E. Lampard, and Richard F. Muth, *Regions, Resources, and Economic Growth* (Baltimore: Johns Hopkins Press, 1960).

[19] To some extent, Stein and Schupack use this approach in making regional projections of the machinery industry. They assume that private investments will take place ". . . where the rate of return on additional capital invested is expected to be greatest. Future expectations are largely based upon past experience. Firms are therefore expected to grow most rapidly in the areas which, in the recent past, have yielded the highest rates of return in a given industry. Experience will indicate whether the businessman was correct in his expectation of future rates of return on investment. If, on the basis of experience, the rate of return on new investment is lower in area *X* than in area *Y*, expectations will be revised in the light of this information. Future investment will be directed towards area *Y* rather than towards area *X*. Expectations are constantly revised in the light of experience. But expectations are heavily based upon recent past experience."

In fact, they found a significant positive relationship between relative rates of return on investment and relative rates of growth of employment, from 1947 to 1954 (Jerome L. Stein and Mark B. Schupack, *op. cit.*, pp. 1-2).

[20] Walter Isard, Eugene W. Schooler, and Thomas Vietorisz, *Industrial Complex Analysis and Regional Development* (Cambridge and New York: Technology Press and Wiley, 1959).

integrated, i.e., takes the form of a regional industrial complex. It defines an advantage as existing if the net revenue of the contemplated complex is greater than that of a similar, efficiently sited and operated complex located in another region.

No matter which over-all approach will prove most powerful for an analysis of a region's assets and relative economic advantage, two conclusions clearly emerge. The analysis will require a huge amount of data, which, fortunately, in an age of high-speed electronic computers can and will be successfully processed. Furthermore, it will be of paramount importance that data needs be clearly defined and that pertinent data be collected in a uniform and consistent manner, preferably by a central body. More careful thinking will be needed before an agreement can be reached on the basic units, e.g., county, for which certain data will be collected. It already is clear that the unit will vary with the particular type of information. For example, data on water, electricity, and industrial fuel, their rates and supply, often suffice on a metropolitan area basis. At the same time, data on local public services and taxes often are needed for specific governmental units.

Recent experience of regional studies has shown that much duplication and waste can be avoided and that improved data can be obtained when information on external forces is centrally collected by a highly qualified research group. Such a central body, specializing in this activity, can then make its findings available to the various regional studies. Likewise, it might be possible to concentrate the industry studies and delegate them to competent centers.

### *Regional Flow Analysis*

Regional flow analysis, in the form of regional input-output analysis, can contribute to assessing the impact of a given internal or external stimulus on per capita income, basic employment stability, net social benefits, and economic growth. This assumes that the initial stimulus has produced known final demand and technical coefficient changes. In many cases the answers will not directly flow from the regional input-output analysis. Instead, a variety of side calculations will have to be made. Wherever appropriate, they will be explored in the following pages.

Before detailed input-output analyses are presented, it is useful to remember some of the conditions that must be met before the analysis is applicable. The following are the most decisive ones:

a) During the period under consideration the region's industrial structure and technology, as well as the local purchase-import mix of inputs, should remain relatively stable.
b) Production functions should be about linear.
c) Average and marginal producers should have about similar technical coefficients.
d) Input-output tables should not be burdened by large-scale aggregation.

More work is needed before a clearer appraisal is possible as to the extent to which these assumptions are likely to be met in regional interaction analysis. Furthermore, since households play such an important role and their inputs as well as outputs are likely to differ depending upon income and occupation, a breakdown along those lines should be considered.

For convenience of presentation, the flow analysis will be presented first in relation to economic growth impact, to be followed by income, stability, and net social benefit impacts.

### Growth Impact Analysis

An evaluation of the growth impact of a given initial stimulus, once it has been translated into the appropriate final demand and technical coefficient changes, if any, can be carried out by using total output, income, or employment as measurements.

To start with, estimation of output impact will be explored. Clearly, the full impact goes beyond the final demand increase that can be traced directly to the initial stimulus. Instead, the increase (or decrease) in a local industrial sector's output, produced by a change in its final demand, unleashes forces which affect the output of all those local industrial sectors which directly or indirectly supply this sector. The input-output flow model makes possible the tracing and evaluation of this chain reaction.

A rather simple and somewhat unrealistic Model I visualizes changes in final demand to lead to production adjustments not only in the specific local sector but also in all those which are directly or indirectly linked to it. However, it assumes that neither consumer expenditures nor investment expenditures for new plant and equipment are affected.

Model II makes allowance also for consumer expenditure adjustments, which come about because of output changes and which lead to a chain reaction of interindustry reactions on income, output, and, once more, on consumer expenditures. It assumes a linear and homogeneous income consumption function.

Model III resembles Model II in every respect except that it incorporates a curvilinear income-consumption function. By all standards this last model is the most realistic.

The three models differ in their implementation, which takes the following forms:

*Model I.* The inverse matrix $(\mathbf{I}-\mathbf{A})^{-1}$ is computed with household and local government sectors part of final demand. It is multiplied by final demand, and the product is horizontally summed to yield an estimate of directly and indirectly stimulated local industry output.

*Model II.* The same steps as above are taken except that the inverse matrix is computed with household and local government sectors part of the endogenous segment of the regional economy.

*Model III.* Income-consumption functions are derived and output reactions are calculated in an iterative fashion.[21] In this man-

[21] Roland Artle in his formal comments took exception to suggestions that changes be made in the matrices during the process of iteration. He stated: "What the type of input-output model, that Werner Hirsch suggests, will give us is really far from 'stages,' 'decision-making processes,' 'actions,' etc. It is in a way as if we took a snapshot of Stage I and another snapshot of Stage IV. But we have only limited information about Stage IV; that snapshot is only partly filled up, in a skeleton way as it were. What we can do, thanks to the input-output techniques, is to fill up the rest of the picture. Input-output is a very fine technique per se. But we shall not try to oversell it by describ-

ner, it is possible to reflect the fact that a final demand change produces direct and indirect income changes which in turn lead to consumption changes and to Round Two of production, income and consumption changes, etc. The process can be traced through three, four, or more iterations.[22]

Instead of measuring growth in physical output terms, an income or employment measure can be applied. Again, for example, all three models can be used to estimate total annual income changes. Model I can be implemented by computing the inverse matrix after household and local government sectors have been moved into final demand. The direct and indirect income change per dollar of final demand is obtained by multiplying each figure of a given row by appropriate household coefficients and summing the products.

The direct, indirect, and induced income changes per dollar change of final demand, assuming a linear and homogeneous in-

---

ing it in terms which it cannot sustain.

"This leads to the following comment. Let us use the following denotations:

$I$ = an identity matrix

$A$ = a matrix of coefficients

$(I\text{-}A)^{-1}$ = the inverse of the $(I\text{-}A)$ matrix

"Due to the special properties of Leontief matrices, we know that we can write the inverse in the form of a power series which will converge very nicely. That is, we can write

$$(I\text{-}A)^{-1} = I + A + A^2 + A^3 + \ldots\ldots$$

"Werner Hirsch now suggests making changes in the matrices as we move along in the process of iterating. I would like to ask permission to disagree very strongly about this procedure. The fact that we are now able to arrange data in matrix form and operate on them according to the rules of matrix algebra gives us new insight. To me, this is real scientific progress, but it seems that we destroy this by interfering quite subjectively with a purely mathematical process."

In contrast to this position is that of Werner Hochwald, who raises the basic question of what constitutes a viable scientific procedure in an operational accounting model. He argues, and the author tends to agree with this position, that: "The very purpose of an accounting model is to facilitate such manipulations which are perfectly 'legitimate' as long as they are applied consistently and accepted by general accounting conventions. Such an interpretation broadens the confines of an 'operational' model and shifts the dividing line between 'scientific' procedures and merely suggestive comments in economic analysis."

Altogether, judicious and consistent interference with the inverse matrix calculations appears proper and is best judged in terms of its usefulness.

[22] For an example, see Frederick T. Moore and James W. Petersen, "Regional Analysis: An Interindustry Model of Utah," *The Review of Economics and Statistics,* 37 (November 1955), 368-83.

come-consumption function, i.e., Model II, are found in the household column of the inverted matrix with households and local government in the endogenous segment.

The very same iterative steps which were taken to estimate the output of Model III produce also estimates of total annual income consistent with Model III.

Next, a few words will be said about employment estimates. Model I is implemented by using the inverted matrix, with household and local government sector parts of final demand, and reading off it the direct and indirect effects of a given final demand change on the production requirements of each of the region's industries. The employment change associated with these changes in production is estimated by inserting each production change into an appropriate employment-production function. Both for Models II and III, employment estimates are obtained in an iterative fashion, similar to the procedure followed to estimate output and income.

### Per Capita Income Impact

For convenience sake, a somewhat special per capita income concept will be used, namely, per man-year income. The reason will become apparent promptly.

Once the total income and employment impact of a change in final demand has been estimated, dividing the first by the second produces per man-year income estimates. Deflating these data by the U. S. Bureau of Labor Statistics' Consumer Price Index results in an estimate of per capita real income.

### Basic Employment Stability Impact

As was stated before, a region's basic employment stability is a direct function of four main variables: (1) importance of the exogenous segment industries; (2) nature of each of the exogenous segment industries; (3) relative importance of output going to exogenous segment; and (4) magnitude of employment multipliers of exogenous segment industries. In addition, attention must be paid to the duration of the cycle, especially the length of the downturn.

Clearly, major final demand increases coming from an industry notorious for the volatility of its demand and associated with a very high employment multiplier, will make the region's economy basically more unstable, and vice versa. The unstabilizing effect of such additions are particularly serious in such metropolitan areas as Detroit and Pittsburgh, where a very high percentage of manufacturing employment is in the durable goods industries, notorious for cyclical instability.[23] Because of these various considerations, the average percentage deviations of manufacturing employment from trend for Detroit and Pittsburgh during 1929-57 are greater than for any other major metropolitan area in the United States. They were 14.6 and 10.3, respectively, while the next largest was Chicago with 9.7.

### Net Social Benefits Impact

At this stage of development it is impossible to do more than examine one or two methods which might help shed light on some crucial elements of net social benefit analysis.

For simplicity's sake, the analysis once more will consider a case where the initial stimulus has resulted in the instantaneous establishment of a new industry inside the region with a known change in the region's final demand and technical coefficients. In addition, the following major assumptions will be made:

1) The regional economy either is operating below its full employment level or it can draw on outside labor resources who are highly mobile and willing to migrate without bidding up wages inside the region.

[23] For example, in 1950 in Detroit and Pittsburgh, respectively, 86 and 83 per cent of manufacturing employment was in durable goods industries. Furthermore, in Detroit the motor vehicle and equipment industry accounted for 60 per cent of manufacturing employment, while in Pittsburgh the primary metals industry accounted for 48 per cent. (Donald J. Bogue and Evelyn M. Kitagawa, *Suburbanization of Manufacturing Activities within Standard Metropolitan Areas* (Chicago: Scripps Foundation and University of Chicago, 1955), Appendix Table B-1, pp. 144-49.

2) To the extent that taxes are shifted, no major income effects result.

3) Government services are person- and not plant-oriented. In this way, the average person in a region can be looked upon as requiring a relatively fixed amount of local services not only in relation to his family and home, but also in relation to his place of work and sale of goods and private services. Thus, costs to local governments for streets and roads, for instance, will be assumed to be closely related to the number of people in the area. In many respects the number of people reflects the presence of plants. The trucking of goods in and out of plants is assumed to have only a minor bearing on the building of roads, particularly when this use is compared with the use made by people who must travel to their homes, places of work, and shopping areas. The assumption that government services are person-oriented might be permissible when one considers that many industrial establishments, especially the larger ones, furnish their own fire and police protection, rubbish collection, et cetera; some even provide for their own water supply and sewage disposal.

Consistent with these assumptions it is possible to construct models which help account for regional direct, indirect, and income-induced costs incurred by various governments in the region resulting from the new industry. As was stated before, it is obviously more difficult to reflect social costs than accounting costs.

One approach would involve the use of a balanced regional input-output model to generate regional government cost estimates. Revenue could be assumed to equal cost. The three levels of government could either appear in separate columns and rows or could be consolidated into a single column and row. There often is a distinct advantage in having separate columns and rows for each major government service and level of government. In this manner it becomes somewhat easier to adjust accounting data to reflect social costs and also to make estimates of tax shifts to "foreigners."

Insofar as the costs of local government services are concerned, there is one particularly bothersome problem. In 1959, property taxes accounted for about 87 per cent of local government tax receipts and 70 per cent of taxes plus charges.[24] Yet, at least in the short run, an industry's sales volume and property tax payments are correlated only to a minor extent. For example, an extended strike can reduce a company's sales to zero, while its assessed value of real property is unlikely to change. Admittedly, this is an extreme case.

To handle this situation calls for some special steps. One method can be employed in case the input-output function is linear, although not homogeneous. If there is but one observation, i.e., for the base year, a function of inputs to outputs can be created by drawing a straight line through that observation and the origin. (See Figure 3.) If it is known that the input is a less-than-

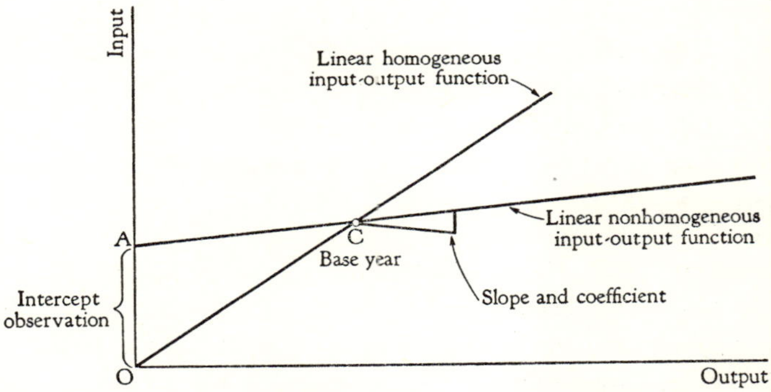


*Figure 3.* Linear nonhomogeneous input-output function.

proportional function of the output (but still linear), this information can be profitably used. Such a function is represented by inserting its slope as a coefficient in the appropriate cell of the matrix, and by stipulating the intercept *OA* as a component of final demand in the row of the input.

A second model will be sketched out in the hope that it will permit more reliable estimates of social costs for government services. A local employment multiplier is applied to help estimate

[24] U. S. Bureau of the Census, *Governmental Finances in 1959* (Washington, 1960), p. 16.

the number of people who will require government services. Since public education expenditures loom very large, separate estimates of population and school population appear in order. With the aid of estimates of the number of those who require services and service costs, total cost of government services can be estimated. Here too, in order to shed maximum light on social cost aspects and tax shifts, these estimates should be made separately for different services and levels of government.

No doubt the preceding pages indicate great conceptual and empirical difficulties in estimating social costs of government services. On the benefit side, matters are even more complex. At the beginning, it might be useful to study cases where the initial stimulus is likely to affect social benefits of government services rendered to the people of a region in a minor way only.

## *IV. SUMMARY*

There can be no doubt that even in a piecemeal implementation of the regional interaction model, relative regional economic advantage analysis provides a major bottleneck at present. Certain elements of regional flow analysis have been implemented in relation to employment, per man-year income, and employment stability impacts, with the St. Louis metropolitan area of 1955 constituting a case study. The results are encouraging if for no other reason than that, depending on which industry was stipulated to incur a specified final demand increase, impacts differed greatly and decisively. With St. Louis manufacturing divided into sixteen major industries, the direct and indirect employment impact of the industry with the largest impact was about ten times as large as that at the other end of the spectrum. The per man-year income differences were not quite as large—the ratio was 3:1.[25] Work to estimate final fiscal residuum is under way.

[25] Werner Z. Hirsch, "An Application of Area Input-Output Analysis," *Papers and Proceedings of the Regional Science Association*, 5 (1959), 79-92. See also, "Interindustry Relations of a Metropolitan Area," *Review of Economics and Statistics*, 41 (November 1959), 360-69.

As for the future, two items loom especially large. At present, on the national level, income and product accounts are more generally available and more highly developed than any other system of accounts. Already today personal income data are available by states, and soon those for major metropolitan areas will be added. While the conceptual problems of regional income and product accounts are numerous, more attention will have to be given to them as well as to their possible integration with regional input-output accounts. At this point, it appears that the difficulties of balancing regional income and product accounts in a meaningful manner are especially great. Careful attention will have to be directed to an exploration of working only with regional product accounts and integrating them with regional input-output accounts. By dropping regional income accounts, relatively little pertinent information might be lost. On the other hand, product accounts are very important since they are closely related to and shed light on regional resources and their development.

The possibilities of further developing and implementing a more general structure for regional analysis are intriguing as well as challenging. There can be no doubt that a determined and intensive two-pronged attack is required. Many conceptual issues must be pursued with imagination and vigor. Particularly intriguing are the possibilities of developing accounts for noneconomic aspects of regions. And yet, even before all conceptual issues have been solved, an all-out effort must be made to clear the way for the development of pertinent data needed to ultimately implement the conceptual framework. The data-collection phase needs to be centralized, most likely within the federal government.

Lead times in procuring data, even under the best of circumstances, are very long. For example, for census data the lead time is four years at minimum. Thus, as soon as some progress has been made in the concept area, an immediate inquiry into the allied data problems must be initiated. Any other approach will greatly delay the day when a general structure for regional economic analysis will become operational.

## COMMENT

*Roland Artle*

*University of California at Los Angeles*

. . . . When anyone sets out to write a report on research, his main concern should be to write for the next man, the one who is to proceed along the same line of inquiry. Professor Hirsch's paper fares very well in this respect.

. . . . I believe that what is needed mostly today in economics is not to build upwards, toward general systems and theories, but rather to build downwards, toward some understanding of the individual decision-making units and their actions and interaction.

. . . . The conceptual framework used by Hirsch is of a non-optimizing type. The "inputs" are local public policies and "external forces." The "outputs" are "the dimensions of the region's health and well-being." They are five in number and they form a vector, if you like. Werner Hirsch stops here; he does not transform this vector into a scalar by means of multiplying it into another vector that would consist of some sort of value judgments.

I have a general remark about the research strategy here. It seems to me that Werner Hirsch has spent too little time on the relations between the "inputs" and the "regional interaction model." For instance, there is quite a distance between changes in land-use regulation or in local taxes and the fact that a new plant will be located in the area concerned. In manipulating the model, Hirsch simply *assumes* that a new plant is located in the area, without going into the connections between this event and the "inputs."

. . . . What Werner Hirsch's suggested type of input-output model will give us is really far from "stages," "decision-making processes," "actions," etc. It is in a way as if we took a snapshot of Stage I and another snapshot of Stage IV. But we have only limited information about Stage IV; that snapshot is only partly filled up, in a skeleton way, as it were. What we can do, thanks to the input-output techniques, is to fill up the rest of the picture. Input-output is a very fine technique *per se,* but we shall not try to oversell it by describing it in terms that it cannot sustain.

Due to the special properties of Leontief matrices we know that we can write the inverse in the form of a power series, which will converge very nicely. That is, we can write

$$(\mathbf{I}-\mathbf{A})^{-1} = \mathbf{I} + \mathbf{A} + \mathbf{A}^2 + \mathbf{A}^3 + \ldots\ldots \text{ where}$$

$\mathbf{I}$ = an identity matrix

$\mathbf{A}$ = a matrix of coefficients

$(\mathbf{I}-\mathbf{A})^{-1}$ = the inverse of the $(\mathbf{I}-\mathbf{A})$ matrix

Werner Hirsch suggests making changes in the matrices as we move along in the process of iterating. I would like to ask permission to disagree very strongly about this procedure. The fact that we are now able to gather and arrange data in matrix form, and operate on them according to the rules of matrix algebra, gives us new insight. To me, this is real scientific progress, but it seems that we destroy this by interfering quite subjectively with a purely mathematical process.

*Harold J. Barnett*

*Wayne State University*

The substantial merit of Professor Hirsch's article is that it provides concepts for analysis of regional economic welfare and growth. His structure of economic variables and parameters provides insight and understanding for some economic relations significant in regional problems. My original comments at the conference concerning the meaning and validity of some of his relations have already served what useful purpose they could, by influencing Professor Hirsch's revised manuscript. In this short note, I shall take another tack. I shall briefly indicate how difficult it is to bring into being a key subassembly—the regional input-output table—in Professor Hirsch's machine.

Input-output relations in his scheme constitute the mechanism by which assumed or actual initial impulses are multiplied and otherwise converted into ultimate economic effects. Three decades of innovation have elapsed since Professor Leontief first "invented" the interindustry table, and for perhaps fifteen years serious academic efforts have been applied to regional input-output tables. We have learned the answers to some important questions.

1) Is Leontief's idea of input-output tables for analysis of economic interaction conceptually important? Yes. Including its operations research offshoots, particularly activity analysis, it is one of only two or three economic analysis breakthroughs during the past generation. Economic principles have been expanded and amplified directly or indirectly because of this innovation, and new areas of applied analysis opened up.

2) Is it possible to construct input-output tables, and regional tables in particular? Yes. In this country, we have the 1947 Bureau of Labor Statistics table, and, at the regional level, single-year tables for St. Louis (Werner Hirsch) and Kalamazoo (National Planning Association and the Upjohn Institute). Abroad, there are input-output tables for a variety of countries, and at least one for a region (Artle's Stockholm table).

3) Have U. S. national or regional tables directly and significantly entered into definitive economic planning or programming activities? No. So far as I know, the U. S. table has not been used in national planning, and the few regional tables compiled from local data have not been used in regional economic planning. None of the large-scale regional planning studies (e.g., New York, Pittsburgh, Cleveland, Penn-Jersey, Southeastern Michigan) has compiled or used local input-output tables. (But one large study, Chicago Area Transport study, "borrowed" the fourteen-year-old national table of coefficients for use in its regional planning study.)

In summary, despite the facts (a) that input-output tables are conceptually the best means we know for tracing effects of an impulse on an economy, (b) that we know how to construct tables, and (c) that there have been a number of large regional planning studies, nevertheless regional input-output tables have not been constructed and used in definitive regional planning and decisions.

The reason is that construction of input-output tables adequate for applied regional decisions is difficult, expensive, and time-consuming. It is not simply *an* input-output table that is needed for regional planning. The table must be appropriate to the major regional problems—those associated with long-term regional growth. The coefficients must be adjustable for the feature of trend and for the more violent changes due to regional exports and imports shifts. In turn, this means that a succession of input-output tables must be constructed if there is to be any chance at all of making estimates of the trend and external trade influences on coefficients. And, therefore, a rough estimate is that a capable empirical research group and a source of funds have to agree to invest roughly ten years and a million dollars or more in model formulation and empirical research capital formation, merely in order to provide a *basis* for definitive use of input-output analysis in regional planning. Beyond the initial investment time and cost is the need for continuing replacement investment to keep the accounts up to date. The very rough guides we have on empirical research investment are the time and cost of bringing national income accounts to the stage where they could be used in national decision-making; the time and expense to move from W. C. Mitchell's original price index proposals to the point where they became tools in national and other decisions; the time and money investment by the National Bureau of Economic Research to create business cycle data suitable for responsible public decisions; the investment in the 1947 input-output table; the investment by RAND and other military groups to compile military cost factors; the investment in the Federal Reserve Board index of industrial production to bring this to a state reliable enough for public use.

We thus have two questions. First, what kinds of local organizations have the talent and long-term horizons to engage in such economic data innovation and capital formation? Second, what kind of organizations have the funds for such farsighted investment? The organizations charged with the social function, state and local governments and quasi-governments, have progressed since Lincoln Steffens wrote his indictment of them—but they haven't improved sufficiently either to be able to attract suitable permanent staff and to permit them adequate freedom to function, or to make the long-term required investments. Universities have the talent, appropriately long view, and interest. But they lack the funds. Local foundations sometimes have talent, interest, and long view, but their resources are usually small relative to the need. Major national foundations have funds. But, except for rare occasions, such as when they underwrote the initial periods of the National Bureau of Economic Research and the Cowles Foundation, they seem to have too short a time horizon and/or inadequate understanding of the ultimate usefulness and cost of such expensive, long lead-time "basic research" and innovation as (in this case) the collection and processing of data for industry interaction models at the regional level.

# 2

# RELATIVE REGIONAL ECONOMIC GROWTH: AN APPROACH TO REGIONAL ACCOUNTS

*Harvey S. Perloff*[1]

*Resources for the Future, Inc.*

MANY KINDS of decisions made by both private and public units depend on analyses of past growth and estimates of future growth of subnational areas. A significant proportion of all the capital investment and service expenditure decisions made by state and local governments, as well as capital investment and production decisions of private industry, is concerned with estimates of regional changes in population, in the volume of economic activities, and in the levels of living—i.e., in the key elements of relative regional growth. Because of the importance of such estimates, it seems logical that they have a central place in any system of formally organized regional economic accounts.

[1] A central purpose of this paper is to try to provide an overview highlighting the relationship between research in the field of regional economic growth and industrial location and work in the field of regional accounts. Both fields are in a period of rapid development so that one cannot hope to do more than suggest potentially fruitful lines of approach. Essentially, the paper deals with the "external" forces as they influence the growth and development of individual regions and the focus generally is "from the outside in."

The author wants to express his appreciation to Jesse Burkhead and William Vickrey for their constructive suggestions made as discussants at the St. Louis Conference and also to Werner Z. Hirsch and Lowdon Wingo, Jr. for their thorough and thoughtful review of the manuscript.

## *I. A REGIONAL-GROWTH FRAMEWORK AND A TWO-LEVEL SYSTEM OF ACCOUNTS*

A focus on relative regional growth and on the factors behind it, in fact, provides a useful way of viewing the requirements for regional accounts as well as the potentialities for constructing such accounts. The regional-growth–industrial-location framework of analysis provides a relatively comprehensive and consistent context for understanding and evaluating subnational economic and physical development. This framework has the additional advantage of tying regional accounts to the mainstream of theoretical and applied work in economics. By its very nature, such a framework tends to focus attention directly on meaningful economic relationships, rather than turn in on itself to provide a neat set of bookkeeping arrangements for their own sake. As will be shown in some detail later, industrial location and localization—using these terms to refer to the location of all economic activities, including production and consumption activities—furnish valuable reference points for all types of analyses. This extends to questions concerned with the geographic form of, and physical relationships in, the metropolitan region (one of the most important and difficult elements in the planning and development of local communities).[2]

The regional-growth–industrial-location framework suggests that a system of regional accounts would have to be built up at two levels: (1) at the national level covering all the regions (i.e., those specifically designated for accounts purposes), and (2) within the individual regions, so that adequate detail can be achieved and due consideration given to the special circumstances and special problems of each community. The information provided at each of these levels would be useful to the other, and would tend to strengthen the other. Over time, individual "consumers" of regional

[2] This is not to suggest that the economic and physical features are the only ones of importance on the subnational scene. Social, political, and other considerations are, of course, centrally involved in many types of decisions made in states and localities. The key point is that the economic and physical features tend to be the *direct* foci of the great bulk of public and private decisions involving information on aggregates; thus, they provide a useful *starting point* for a regional accounts effort, no matter how encompassing the latter turns out to be over time.

accounts may be expected to learn how to use each of these types to advantage.

A complete coverage at the national level is needed for federal government decision-making as well as for decisions made by private enterprises whose activities extend over more than one community or state. It is equally necessary for the analysis of problems and for decisions made within any particular region. The development of any region can adequately be understood only within the context of national economic and social development; the relationship between the parts and the whole is always a critical element in regional analysis. In addition, comparative data of the type provided by nationwide coverage are a powerful tool in regional analysis. The need for nationally prepared, across-the-board estimates is especially evident with regard to growth projections (and, as already indicated, projections are crucial in subnational decision-making). Growth projections developed for a single local unit often tend to be exaggerated; it is only too easy to claim a larger share of the anticipated national total increase than is reasonable. Projections for the whole nation provide a corrective for this. And, of course, they provide a tremendous saving through the "economies of scale."

A system of regional accounts built up at two levels would be particularly useful in application if both employed the same geographic building blocks. Since various kinds of public and private decisions necessarily cover a variety of types of subnational units (from the small county to the multi-state region in various combinations), the main possibility for general usefulness lies in providing information for sensible and flexible data "building blocks." If we focus attention on the possibilities for the decade of the 1960's, a strong case, I believe, can be made for using the states and the Census-defined Standard Metropolitan Statistical Areas as the basic data units for regional accounts. Given data for these units, some breakdowns will be possible for the sub-SMSA units and for the nonmetropolitan areas of the state, by way of allocation techniques. Even units as large as the SMSA's will provide plenty of headaches—for example, with regard to the disclosure problem in manufacturing industries—but these are highly significant units for decision-making and deserve particular attention in the construction of regional accounts.

What types of data and what kinds of accounts does the logic of the regional economic growth model suggest? While it might be interesting to follow through this logic without any regard to feasibility, it seems preferable to discipline the approach by weighing the possibilities of obtaining the necessary data within some middle-range period, say, a decade. Within such a period we can look forward to data and accounts progress in a number of directions.

Extremely important in this regard is the work under way within the Office of Business Economics of the U. S. Department of Commerce to prepare interindustry sales and purchases (input-output) data for 1958, integrated conceptually and statistically with the national income and product accounts. When completed, the basic estimates "will be maintained on as current a basis as the available data permit." What is particularly significant as regards future possibilities for regional accounts is the fact that the sales-purchases information will involve a deconsolidation of the national income and product account *along industry lines.* As explained by the OBE, income and product accounts for the various industries will be established. The product (credit) side will show, in addition to sales to final users and inventory change, the sales of goods to each intermediate industry. The debit side will show, in addition to gross product originating in each industry, the purchases of intermediate goods by each industry. When set up in matrix form, the final demand columns will reflect the expenditures for gross national product; the value added rows will show gross product originating on an industry basis. This information will provide an extremely valuable framework for regional estimates and projections, as well as for regionally constructed input-output tables.

Another important development is the fact that we can expect to have personal income data on a regional basis for the major metropolitan areas as well as for the states.

Most of all, looking ahead, we might hope that to an increasing extent key national income and product and sales-purchases aggregates—or, more specifically, the major "control totals"—will be developed centrally for states and metropolitan areas and that there will be provided at the national level those state and metropolitan tabulations of data which can be obtained by normal data processing methods in the course of developing the national aggregates (such as regional tabulations of the data from the Economic Censuses and from the Internal Revenue Service).[3]

We can take it for granted that over time there will be continuing play back and forth between improved concepts and tools for regional economic analysis and the strengthening and improvement of regional accounts. What is involved at this stage is necessarily an early effort to develop the logic of a systematic approach to the design of regional accounts, starting from where we are in our understanding of regional economic growth. This suggests, first, determining the needed measures of economic growth (or change) and then a look at what would seem to be the key analytical variables.

## *II. MEASURES OF CHANGE*

How is regional growth (or change) to be measured? It can be assumed that at least four types of socio-economic change are significant for decisions made within or about metropolitan communities and states.

1) Changes in the *volume* of economic activities and in population.
2) Changes in *welfare.*
3) Changes in *productivity.*
4) Changes in economic *stability.*

Several tests seem logical in choosing indices of change. One is inherent interest in the index itself, as in the case, say, with total

[3] Richard Ruggles and Nancy Ruggles, "Some Relationships Between National Accounts and Regional Accounts," *Proceedings,* American Statistical Association, 1959, pp. 278-81.

population or per capita income. A second is its usefulness as an analytical variable. And a third is the possibility (or ease) of obtaining the data.

Employing such criteria, a case can be made for measuring changes in *volume* by way of the following items: population, labor force, number employed, total personal income, and total value added by various industrial sectors. The key *welfare* index is normally taken to be per capita income. However, a more meaningful picture could be provided if in addition there are data on distribution of income by family income classes, on wage levels, on number of unemployed, and on net fiscal residuum (as suggested by Hirsch). *Stability* might be represented by changes from a trend line—monthly, quarterly, or annually—in employment and unemployment and in value added. Changes in the productive *efficiency* of the regional economy can be measured in terms of changes in the value of output of goods and services (value added) per worker. All these elements have been discussed at great length in the literature[4] and it is useful here only to refer to them as background for the discussion which follows.

In terms of regional accounts, such a listing of measures of change, tested against currently available data, suggests several "new" requirements: (1) the need for value added estimates for states and metropolitan regions for all major economic activities on a periodic basis; (2) the need to place Employment Security data on employment and wages on a consistent basis (particularly with regard to industry and occupational classification) and regularly publish the data for all states and metropolitan regions; and (3) the early need for detailed periodic information on family

[4] Among other sources, see *Regional Income*, Vol. 21 of Studies in Income and Wealth, National Bureau of Economic Research (Princeton: Princeton University Press, 1957). Suggestive materials on the measurement of economic stability are provided by George H. Borts, "Regional Cycles of Manufacturing Employment in the United States," *Journal of the American Statistical Association*, 55 (March 1960), 151-211. On the relationship between these indices of change and the designation of social objectives or goals, see Werner Z. Hirsch's paper for this Conference.

income through Internal Revenue Service data and other sources.

## *III. ANALYTICAL VARIABLES: A REGIONAL GROWTH MODEL*

If we had all of these indices for all the states and metropolitan regions of the nation over a number of years, we would certainly have a great deal of valuable information concerning socio-economic change. However, it is evident that more would be needed for analytical and projection purposes. More than anything else, we are concerned with having—in an effectively organized way—the raw materials for analysis of the problems that concern the public and private decision-makers, and, of course, the general public.

We can refer to current knowledge of the factors behind regional economic growth to identify key analytical variables, and therefore data requirements. Regional economic growth is clearly a highly complex phenomenon. However, certain central features can be identified.[5]

A useful procedure is to visualize the requirements for making regional growth projections (assuming a comprehensive, eclectic approach), since this *directly* relates regional growth theory to the accounts question. Taking the projected national demand for various products and services as a starting point and translating this into total production by industry classes (gross value, value added, and employment), the problem is to allocate supply among the various regions of the nation. In turn, it would be necessary to allocate purchases on a regional basis—that is, to identify on a geographic basis the markets for the various goods and services.

[5] The analysis here draws on materials presented in Harvey S. Perloff, Edgar S. Dunn, Jr., Eric E. Lampard, and Richard F. Muth, *Regions, Resources, and Economic Growth* (Baltimore: Johns Hopkins Press, 1960). This book, in turn, draws on the work of many analysts, including Alfred Weber, J. H. von Thünen, August Losch, E. M. Hoover, Jr., Walter Isard, W. H. Dean, H. A. Innis, D. E. North, Colin Clark, Bertil Ohlin, and others.

The interrelations of the two would have to be worked out, of course, by successive approximations. Relative regional growth (in volume) would be the reflection of the relative proportion of the total increases in supply and demand accounted for by each of the regions (but with different figures expected for increases in value added and in employment, because of the regional variation in labor productivity).

Production and consumption data (with income data serving as a bridge between the two) have to be played back and forth because each so directly influences the other, and these are, in fact, the logical organizing foci for most of the data. The subcategories are, then, the aggregates which highlight the more meaningful relationships. One of these arises from the significant differences between the "export" industries and the derived or residentiary industries. This distinction can be symbolized by comparing the role of the steel industry as against the role of retail trade in the growth of the Pittsburgh metropolitan region, or comparing the place of the aircraft industry as against that of personal services in the development of California. In other words, the major export industries (which include the federal government activities) are seen to have a special "lead" role in regional growth and to be "autonomous" from the standpoint of any particular region. The employment and income generated by the export industries provide the main base on which the so-called derived or residentiary industries will be built. This relationship is not uniform everywhere or over time.

The extent to which the derived industries will be built up will depend on various factors—such as the types of forward and backward linkages provided by the export industries, the service requirements of these industries (e.g., the nature of the freight shipments and the volume of business services used), the wage and salary levels set by the export industries, and the like. These are analytically important and have to be considered specifically in making the regional projections. But while the growth of the derived industries is by no means automatic, it tends to take place within a fairly circumscribed range. It is this feature, rather than any inherent sense of relative economic importance, that makes the export-derived industry distinction analytically useful.[6]

On the consumption side, the important distinction is between final consumption and intermediate consumption. The former is closely associated with the derived industries, but not entirely; most of the export industries that produce final goods and services will sell at least part of their total product locally. In fact, the size of this local sale could be an important factor in the location decisions of certain of the industries. In the case of the goods-producing export industries, however, the important element is the purchase of intermediate goods—since the relative costs of acquiring (and selling) such goods is weighed in location and production decisions.

In making the regional projections, then, the starting point is a determination of the probable location and production of the "export" industries (specifically, those serving markets that extend beyond one state or one SMSA). This is not so much a matter of industries literally moving from one part of the country to another, within the projected period, as it is a question of where the new capacity will be built and the increases in production will take place. The answer has to be determined on the basis of *relative regional advantage* with regard to the quite specific requirements of each of the major industries. These relative advantages can be analyzed in terms of the classic factors of accessibility of natural resources, quality and quantity of labor (including elasticity of supply of labor), and size of markets (together with other factors related to economies of scale), as well as the somewhat newer factor of the relative desirability of the working and living environment. These, taken together, will determine which region or regions would seem to offer the best over-all economic (profit-earning and/or capital-preserving) position for each of the major industries. Where one consideration is dominant, as is often the case, the industry can be characterized as resource-oriented, labor-oriented, market-oriented, or amenity-oriented.

[6] It should be noted that these categories are treated here as useful "organizing" classes in an industry-by-industry approach to national-regional projections and not as an automatic "basic-nonbasic" technique for regional projection. For the limitations of the latter, see Ralph W. Pfouts (Ed.), *The Techniques of Urban Economic Analysis* (West Trenton, N.J.: Chandler-Davis Publishing Co., 1960).

All this suggests that both the requirements of the major industries and the relative regional advantages with regard to these requirements have to be examined separately and in some detail. This calls for a variety of data, and these can be usefully organized in three major categories: data centering on (1) manpower or labor characteristics; (2) input-output and income-outgo; and (3) stock (assets or wealth). The first are data concerned with the human resources facet; the second, the intermediate-flows facet; and the third, the nonhuman capital facet. The key factors involved in analyzing *relative regional advantage* can be subsumed under these categories.

On the production side, the manpower information should throw light not only on numbers but on characteristics (such as sex, race, etc.) and on occupational and skill groupings as well as on levels of labor payments. The flows data, covering the interindustry purchases and sales, can provide insights on relative transportation costs and linkage advantages or disadvantages, as well as on growth of markets, investments and returns on investment. The stock category ideally should provide information on resources and amenities, as well as on output capacity and value of assets. The difficulty of obtaining information within this last category is obvious, but its importance should be equally obvious.

On the consumption side, the flows are of course of particular interest. The income and expenditures of households, as well as the purchases of firms, are needed to provide a measure of the changing levels of final markets and of the intermediate markets. The totals are important because of the increasing market orientation of industry generally and because of the significance in regional growth of the achievement of higher regional market "threshold" levels. The latter refers to increases in market size adequate to attract a whole new set of industries—say, meat packing or automobile assembly—that find they can now achieve adequate economies of scale by production within a given region to satisfy regional final or intermediate demand.

Since we are going through the exercise of visualizing the requirements for making regional growth projections as a way of highlighting the requirements for regional accounts, it is useful to note how the analytically significant data might be organized for making the regional projections. The problem is one of breaking down national projected population and economic totals (employment, output, income, purchases, etc.) into regional components—i.e., totals for the states and metropolitan regions. To do this, it is necessary to look both backward and forward, the historical series providing the first approximations from which departures can be made by way of analysis of changing conditions and their impact on the key relationships.[7] The initial question is one of relating the regional totals to the national totals—the parts to the whole—in a meaningful way. This can be done by translating the regional time series into "share" and "shift" terms so that it becomes easier to relate the parts to each other as well as to the whole—that is, who has gained at whose expense? These data could also be used in analyzing hierarchical relationships where they exist, such as any ties between the growth of the metropolis and the growth of the state or multi-state region.[8]

The data showing year by year what proportion or "share" of the national totals are to be found within each of the regions, not only provide valuable across-the-board information, but also provide a useful type of discipline. They indicate clearly that, given certain assumptions about national growth, unusually rapid gains in certain parts of the country must be offset by stagnation or decline in other parts of the country; and then one immediately has to answer the question of where these declines as well as gains are

[7] In his discussion paper, Jesse Burkhead highlighted the dangers inherent in any automatic projection of relationships from historical data, particularly given the problems associated with the instability of production coefficients and trading coefficients and the likelihood of the coefficients being more unstable in the future than in the past (stemming from the impact of the federal highway program, the appearance in certain regions of blocks to development such as water shortages, etc.).

[8] As a matter of fact, metropolitan areas within the various large multi-state regions show similar over-all growth patterns in many cases. See Perloff, Dunn, Lampard, Muth, *op. cit.*, p. 44. See also O. D. Duncan, *et al.*, *Metropolis and Region* (Baltimore: Johns Hopkins Press, 1960).

to take place.[9] In the same way, the "shift" data for the major measurement items—for example, for total population, employment, income, and value added—serve to relate each region's growth to the national averages. Here one can see clearly whether regional changes, even if generally upward, have been greater or less than the national average change. Thus, if a metropolitan region or state has been growing in employment and income faster than the nation as a whole, an "upward shift" is recorded—that is a shift greater than that expected on the basis of the national average—and the question of what is behind this immediately comes to the forefront.[10]

A good starting point to arrive at an understanding of the changes that have taken place as among regions are the shift data in terms of the industry components. Thus we would want to have figures for the shifts in employment and in value added for each of the major export and derived industries for each of the regions. To

[9] The type of information provided by this form of organization of data can be suggested by an example using total personal income data and multi-state regions.

REGIONAL DISTRIBUTION OF TOTAL PERSONAL INCOME, 1900-1957
U.S. income in millions

| | 1900* | 1920*† | 1930 | 1940 | 1950 | 1957 |
|---|---|---|---|---|---|---|
| United States | $15,391 | $69,276 | $76,780 | $78,522 | $225,473 | $345,272 |
| | 100% | 100% | 100% | 100% | 100% | 100% |
| New England | 9.9% | 8.8% | 8.6% | 8.2% | 6.7% | 6.6% |
| Middle Atlantic | 30.8 | 30.2 | 33.3 | 30.5 | 26.4 | 25.5 |
| Great Lakes | 22.4 | 22.2 | 22.6 | 22.7 | 22.5 | 22.5 |
| Southeast | 12.0 | 13.0 | 11.2 | 13.2 | 15.2 | 15.4 |
| Plains | 13.3 | 10.3 | 8.9 | 8.3 | 8.8 | 8.1 |
| Southwest | 3.8 | 5.7 | 4.8 | 5.2 | 6.5 | 6.8 |
| Mountains | 2.5 | 2.5 | 1.9 | 2.0 | 2.2 | 2.2 |
| Far West | 5.3 | 7.4 | 8.8 | 9.9 | 11.7 | 13.0 |

* District of Columbia excluded. † 1919-21 average. Source: Perloff, Dunn, Lampard, Muth, *Regions, Resources, and Economic Growth*, p. 24.

[10] An illustration here might also be useful. The table below shows for 1939-54 the state changes in population relative to the change in population for the nation as a whole for the states that have experienced the greatest

get at what is behind the shifts in relative standing among the regions, a distinction needs to be made between the composition effect—that is, the consequences of a region having a favorable or unfavorable "mix" of the rapid-growth versus the slow-growth industries, as against the differential (or local factor) effect—that is, shifts among regions *within* given industries due to changes in relative regional advantages for these industries. The importance of this distinction stems from the fact that regions can grow *either* because they have industries that throughout the whole nation are enjoying a rapid growth *or* because they are getting a larger and larger proportion of a given industry—whether it is growing or not.[11]

The composition—or rapid-growth industry—effect arises from certain significant national factors, such as changes in taste, relative income elasticities for products, price elasticities, productivity, etc. In making regional projections, a large part of the total effort must necessarily center on analysis of just these factors as they relate to the relative growth or decline of *specific* industries on a *nationwide* basis.

---

departures from the average. All the "upward shifts" add up to 100%, as do all the "downward shifts" taken together. Thus, California had some 44% of the total above-average shifts during the period 1939-54. For some types of information year-by-year shifts need to be calculated.

PRINCIPAL POPULATION SHIFTS AMONG THE STATES, 1939-1954

| Percentage net upward shift | | | Percentage net downward shift | | |
|---|---|---|---|---|---|
| (1) | California | 43.9 | (1) | Pennsylvania | 14.1 |
| (2) | Florida | 11.9 | (2) | New York | 8.7 |
| (3) | Michigan | 7.7 | (3) | Oklahoma | 7.3 |
| (4) | Texas | 6.7 | (4) | Arkansas | 6.3 |
| (5) | Maryland | 4.6 | (5) | Mississippi | 5.7 |
| (6) | Washington | 4.4 | (6) | Missouri | 5.7 |
| (7) | Ohio | 3.8 | (7) | Massachusetts | 5.6 |
| (8) | Arizona | 3.5 | (8) | Kentucky | 5.1 |
| (9) | Oregon | 3.3 | (9) | Iowa | 4.6 |
| (10) | North Carolina | 2.3 | (10) | Alabama | 4.1 |

Source: Perloff, Dunn, Lampard, Muth, *Regions, Resources, and Economic Growth*, Appendix Table H.

[11] For a detailed explanation see Perloff, Dunn, Lampard, Muth, *op. cit.*, Chap. 5. Also Edgar S. Dunn, Jr., "A Statistical and Analytical Technique for Regional Analysis," *Papers and Proceedings of the Regional Science Association*, 6 (1960), 97-112.

The differential effect will require a look at the special local factors as they play themselves out with regard to the requirements of each of the major industries. The point here is that industries, whether growing or declining nationally, will not grow in such a way that each region retains its exact share of the total, but rather that some regions are likely to get a larger and larger share of the total capacity and production over time. Thus each region has to be looked at with regard to its changing relative advantages as to its input and market position for each of the major industries.

It is here that the various manpower, flows, and stock data, as described above, come into play. An especially important focal point is relative marginal productivity of capital among regions and the reasons behind the differentials. It can be assumed that as new capacity within an industry is required due to anticipated increases in demand, such capacity will be built in those regions where the returns on investment can be expected to be largest.[12] Thus, past data on marginal productivity of capital is highly pertinent. However, since relative advantages with regard to inputs and markets are constantly subject to change, the regional analyst has to go through more or less the same kind of analysis that the businessman does in making his decisions with regard to new construction of capacity. That is, the past is a valuable indicator but not the total picture. The new situation as it is evolving must be taken into consideration.

[12] Stein and Schupack have used relative marginal product of capital to predict regional employment. See Jerome L. Stein and Mark B. Schupack, *Forecast of New England Machinery Industry in 1970,* Research Report of the Federal Reserve Bank of Boston, 1970 Projection No. 11, December 1959. Ideally, we would want to have figures on the value of capital assets, on capacity, on replacement and new investment, and on return on investment. An outstanding study employing these measures in projecting employment growth is T. Y. Shen, *New England Manufacturing Industries in 1970,* Research Report of the Federal Reserve Bank of Boston, 1970 Projection No. 1 (no date). The surveys that have been carried out by the Federal Reserve Bank of Boston suggest that rough estimates, at least, are feasible (although just how accurate present estimates are is an open question). In general, it is clear that we need a good many studies on capital, both national and local, to provide a firmer basis for estimates. This is a weak link at the present time and attention needs to be devoted as to how we might obtain the needed data on capital and on investment returns.

Once estimates have been made for the export industries,[13] and the independent population additions, the base has been provided for making volume estimates for the derived industries. Here studies of changes over time in the relationship between the export and derived industries and the relative sizes of derived industries among different regions would be highly suggestive. However, in looking at the future, additional analysis would be called for, to account for forces such as the following: (1) the effect on the construction industry of requirements for new overhead, i.e., social capital as well as for anticipated building of plant and equipment; (2) the generation of unusually active renewal programs within given metropolitan regions; (3) anticipated changes in regional income distribution; (4) the effect on services of an unusual rapid increase in income, if such is anticipated; and (5) the effect of identified "organizational" trends in various industries, such as the decentralization of banking services. A useful distinction can be made between minimum-size derived industry totals (i.e., the per capita volume below which no region falls—or the amount that can be assumed will be added with every increase in population) and the above-minimum volume which is the result of the special conditions of the individual region, as revealed by factors such as those outlined above.[14] Generally, shifts in industry and in population are closely related, but because of the increasing

[13] One of the special categories that deserves particular attention is the localized forward- and backward-linked industries; that is, the industries that cluster in space so that they make up localized industrial complexes. These are industries that tend to obtain their best over-all position by clustering together within a relatively limited space rather than trading over a wide area. For these industries we would need detailed data on sales and purchases within the various regions. The petroleum-petrochemical complex and the various metals complexes are cases in point. For a thorough examination of the locational aspects of one such complex, see Walter Isard, E. W. Schooler, and T. Vietorisz, *Industrial Complex Analysis and Regional Development* (Cambridge and New York: Technology Press and Wiley, 1959).

[14] On this point, see Gunnar Alexandersson, *The Industrial Structure of American Cities* (Lincoln: University of Nebraska Press, 1956); Irving Morrissett, "The Economic Structure of American Cities," *Papers and Proceedings of the Regional Science Association,* 4 (1958) 239-56; and Edward L. Ullman and Michael F. Dacey, "The Minimum Requirements Approach to the Urban Economic Base," *Papers and Proceedings of the Regional Science Association,* 6 (1960) 175-94.

volume of the nonjob-oriented population movement, an independent estimate of the latter is called for. We know that the amenity resources—that special constellation of climate, shore, water, etc., which attracts people in large numbers—are playing an increasing role. Some of the current studies on the older groups in the population contain suggestive data. The trends are very powerful here, particularly with regard to the movements to California, the Southwest, and Florida, so that the highly significant shares of this movement can at least be identified.

Projections normally have to go through a series of successive approximations, and this is true for the regional projections discussed here, particularly for the market-oriented industries and the derived industries generally. For these industries, successive market "threshold" effects have to be estimated to avoid too much piling up of multiplier effects on one side, and to avoid an underestimation of the role of changing scale economies. Here the discipline of employing the "share" and "shift" techniques is particularly valuable.

The application of trend lines to the future from the historical series might be done by way of regression analysis, with a good dose of judgment. The objective would be to come up with future values—possibly set out with ranges—for all of the measurement indices; that is, for regional population, labor force, employment, income, and value added. From the standpoint of the users of the regional projections, however, the associated analytical materials would be at least as useful as the projected measurement indices. This is true whether the focus of interest is on increases in the volume of activities or on changes in income levels, in productivity levels, or in economic stability levels.

## *IV. PROJECTIONS FOR INDIVIDUAL REGIONS*

Anyone preparing growth estimates for an individual region—whether a multi-state region, a state, or a metropolitan region—would obviously seek to relate them to the across-the-board projections (assuming that these were available). Otherwise the task for the individual region is formidable. Given the across-the-board projections, however, the estimates for an individual region can focus on a few special considerations in detail and also test alternative projections as possibly influenced by way of public policy and programs. Here impact analysis, as outlined by Hirsch, would come into play.[15]

In addition to providing greater depth, the individual-region studies (and, particularly, metropolitan region studies) are needed to emphasize another element that does not appear in the across-the-board studies—that is, the "localization" patterns, or where economic activities (production and consumption) take place *within* the metropolitan region, and the relative rates of growth of the various sections of the region.[16]

The localization patterns and localization requirements provide a significant link between the national and "large"-regional location factors and the specific public planning and investment decisions that are normally made within a metropolitan region. Thus, on the *intra*regional scene, the general requirements for a given type of labor supply are translated into the ability to obtain the necessary labor within reasonable commuting distance. The general considerations concerning the costs of collecting materials and shipping products are related to the ability to get industrial sites near rail lines or near main highways, as well as to relative traffic and terminal conditions within the metropolitan region. Special requirements, such as the availability of adequate water, also come into play at the localization level, and of course the amenity considerations have their localization features as well as

[15] Werner Z. Hirsch, "Regional Accounts: Objective and Framework," *Proceedings*, American Statistical Association, 1959, pp. 282-90. See also his valuable series, *Three Studies in the Measurement of the Impact of Metropolitan Growth* (Washington: Resources for the Future, Inc., 1959), mimeo. An important step forward in impact analysis is marked by the National Planning Association study, *Local Impact of Foreign Trade* (Washington: 1960).

[16] The terms "location" and "localization" are not used in any uniform sense in the literature, and only rarely is "localization" separately treated, but it is clearly useful to have separate terms which permit differentiating between the *intra*regional considerations and the broad *inter*regional considerations.

the more general region-wide feature of climate. Clearly, relative regional growth and location considerations extend to the *intra*-regional situation.

On the other side, assumptions about additions to population and about growth of economic activities in a given region carry with them a whole series of concomitants at the localization level. These are, for example, requirements for a certain amount of land and for certain specialized sites[17] (based on the kinds of industries anticipated and the characteristics of the persons who will be manning these industries), and changing requirements for transportation and for a whole host of social overhead items like education and health facilities. Here we find our knowledge particularly limited. We need a great deal more information than we have about the relationship between the growth of various types of industries and land requirements, transportation requirements, water requirements, and social overhead requirements in order to translate the generalized regional projections into specific localization requirements. This type of information is needed to understand what has been happening with regard to both land use and service requirements within the metropolitan regions, as well as what changes can be anticipated on the basis of over-all growth.

## V. *THE LOCALIZATION VARIABLES*

The usefulness of the regional growth-location-localization framework for evolving regional accounts can be highlighted further by noting the extent to which, on the intraregional level, the same types of analytical variables are needed as on the larger region "location" level. This grows out of the fact that on the intraregional level, as well as the other, both the producing units and the consuming units normally seek to optimize their over-all position (or values).

[17] Estimates of land and building use by various types of industry are provided in the *Economic Report 1959*, prepared by the City Planning Commission of Sioux City, Iowa, under the direction of Charles L. Leven.

On the production side—including manufacturing, commerce, government and other service activities—there are the autonomous and derived units. The latter are those tied to specific population or market aggregates (such as nonspecialized retail stores, schools and many other types of public services, and various kinds of private trade and service establishments), and tend to be added in "organic" cell-like fashion. These are the counterparts of the derived industries—as against the export industries—in regional location analysis and, like the latter, tend to vary within relatively limited ranges with given additions in population.

The autonomous units normally seek specialized localization features and, in fact, can be characterized in terms of their "orientation" (depending on how large the item looms in the total costs and/or returns of the firm—as well as how much variation is offered by the different localization opportunities). As a tentative listing of such orientations, the following might be suggested:

1) *Transportation-oriented:* requiring special facilities such as ports or rail lines, or sensitive to high transport costs (e.g., cannot afford to fight heavy in-town traffic).

2) *Large-market-oriented:* drawing on the entire region, usually in the central business district, but also in "regional" shopping centers.

3) *Communications-oriented:* requiring a great deal of face-to-face communications and/or nearby specialized service operations.

4) *External-economies-oriented:* overhead service costs loom large and therefore such firms seek locations where these services are or can be provided at minimum cost, as in industrial estates or central business districts.

5) *Labor-oriented:* emphasis on ability to acquire a given quantity or type of labor at a "reasonable" cost (such as high-skill industries drawn to suburbs or industries requiring many workers drawn to locations along main highways).

6) *Large-site-oriented:* such as coal yards and lumber yards as well as manufacturing enterprises whose technology calls for one-story operations; also hospitals and certain other public facilities.

7) *Water-oriented:* requiring location along a river or at a point where large quantities of water can be acquired at low cost.

The projected localization of anticipated new autonomous units can be worked out by relating the industry requirements (or orientations) to: (1) the environmental features of the region, (2) the estimated "holding capacity" of the different subareas of the region,[18] (3) accessibility (transportation cost), and (4) the costs of land in each of the subareas.

Governmental facilities and activities fall within both the autonomous and the derived categories. For "first approximation" projection purposes—i.e., assuming a continuation of existing governmental policy—the governmental units can be treated in basically the same manner as their private counterparts. It could be expected that such "trend" estimates would be one of the types of information employed in the development of new governmental policy.

The localization of the "derived" units is essentially tied to the relative size and character of the consuming publics in the various subsections of the region. The most important of the derived items are the construction of new housing and the provision of housing services, followed closely by the provision of public services and the construction of public facilities, and followed in turn by the more or less ubiquitous private trade and service establishments.

Several techniques have been developed for estimating the future distribution of households in the metropolitan region. These commonly approach the problem by distributing a given number of new units (new households, dwelling units, or population) among the subareas of the metropolitan region by some explicit rule. Lowdon Wingo, in a current study,[19] classifies the various methods as follows: *ad hoc* allocations, gravity (or potential) models, and economic models.

*Ad hoc* methods are characteristic of current city planning practice and typically begin with the determination of "holding capacities" for the subareas based on such factors as the available land,

[18] Defined in terms of the established zoning and density standards. However, as William Vickrey has suggested, for certain purposes, "holding capacity" and development standards need to be viewed as parametrically variable.

[19] Lowdon Wingo, Jr., *Transportation and Urban Land* (Washington: Resources for the Future, Inc., 1961).

prevailing population densities, and standards of development.[20] For the prediction of future traffic requirements, the Chicago Area Transportation study developed a technique which involves the distribution of the estimated population among a large number of small areas according to their distance from the central business districts and a "percent-of-holding-capacity" function evolved from existing relationships.[21]

The potential or gravity model has been applied to urban spatial organization in various ways. Walter Hansen has developed a potential model which will allocate an increment of households among subareas, given for each subarea its employment, a measure of its accessibility to all other areas, and the amount of "developable" land it contains. [22]

Economic models assume that the spatial pattern of the urban economy results from the "optimum-seeking" activities of firms and individuals, each bidding for his "best" combination of two variables: the amount of land and its location. Both linear programming and continuous equilibrium forms have been explored. Herbert and Stevens have developed a linear programming model to allocate households among a number of subareas in a metropolitan region which would take the journey to work and the household demand for space as the significant relationships to be observed. Allocation would be carried out to maximize some function of location rents, space per household, and accessibility to employment, subject to certain restraints such as the supply of space.[23]

Wingo has developed an equilibrium model for the allocation of households based on two key functions of households—the provision of labor services at specific production sites and the purchase of the services of urban land for shelter space. The model concentrates on how the labor services are organized in space, given

[20] Cf., F. Stuart Chapin, Jr., *Urban Land Use Planning* (New York: Harper & Bros., 1957), pp. 339-54.

[21] John R. Hamburg and Roger L. Creighton, "Predicting Chicago's Land Use Pattern," *Journal of the American Institute of Planners,* 25 (May 1959), 67-72.

[22] Walter G. Hansen, "How Accessibility Shapes Land Use," *Journal of the American Institute of Planners,* 25 (May 1959) 79-91.

[23] John D. Herbert and Benjamin H. Stevens, "A Model for the Distribution of Residential Activity in Urban Areas," *Journal of Regional Science,* 2 (Fall 1960), 21-37.

the characteristics of the transportation system, the spatial arrangement of production, the nature of the labor force, and individual household demands for space. It brings together the supply and demand elements to generate a spatial distribution of location rents and household densities. Thus, the model is a close approach to a localization counterpart of an interregional industrial location model of the type described earlier, dealing with the interrelationship of production and consumption elements in space.[24]

For many types of decisions made in or about metropolitan regions involving a focus on subareas, information about numbers alone, whether of population or households or industries, is not enough. All sorts of detailed stock and flow data for specific areas or spatial configurations within the metropolis are needed. This is particularly true with regard to data on human resources and data on public services, expenditures, and revenues. Examples are decisions touching on plans for building schools and other public facilities, on the provision of voluntary health and welfare services, and on estimates of property tax returns. At the present time in metropolitan communities where this type of subarea data is collected, it is done on a haphazard and not-too-well-organized basis, but in looking ahead to improved regional accounts it is evident that such data and their organization must receive high priority.

A great deal of work needs to be done to clarify just what data should be collected on a regular basis for all subareas, particularly given the high cost of providing a wide range of information for a large number of units covering multiple political jurisdictions.

Three important tasks emerge here. One is the development of a sensible basis for demarcating subareas within the metropolis for accounts purposes. Another is the design of a useful data classification system which can serve the major accounts purposes. And the third is the invention of a scheme for the recording and use of the subarea data so that significant spatial relationships—such as distances from the central business district (or the core of the city) or income-potential and population-potential measures—

[24] Wingo, *op. cit.*

can be recorded and analyzed. The total system should be suited to modern machine techniques of data manipulation.

In general, what would seem to be needed is a dynamic picture of the localization of facilities, persons, and activities within the metropolis. The provision of such data would, clearly, be an enormous undertaking, but given the nature of the decisions which call for this type of information—such as the multi-billion-dollar transportation proposals—the costs involved are likely to be justified. It might well be that within the decade "data centers" will be established in the major metropolitan communities and these could provide the information needs of the various types of decision groups on a cost basis.

## *VI. IMPLICATIONS FOR THE DEVELOPMENT OF REGIONAL ACCOUNTS*

The above examination of the information needed to understand and project "large"-regional and intraregional development would seem to point to a number of conclusions with regard to regional accounts.

1. There are a number of important advantages in seeking to provide information in an accounting framework, stemming from the inherent requirement that an effort be made to achieve:
   a) comprehensiveness—"accounting" for the totality of the items involved (such as all of personal income received);
   b) internal consistency—normally based on the "double-entry" nature of recorded transactions, but permitting other checks as well;
   c) comparability—resulting, among other reasons, from uniformly and consistently defined categories; and
   d) periodicity—providing the data at given known intervals, so that potential users can count on their availability.

These highly significant advantages suggest the extension of the accounting framework to cover as much of the basic data needed for decision-making as possible.

2. Regional analysis calls for data of such a wide variety that a number of accounts would have to be developed, including stock (wealth) as well as flow accounts, to satisfy this need. No one type of account is likely to suffice, no matter how comprehensive. The same kinds of considerations which gave rise to the development of five major types of national economic accounts are at play on the subnational scene. The problems that have to be solved at the subnational level are complex and subject to rapidly changing forces; decision-makers have to be armed with various types of information and there are decided gains if the essential data can be organized into accounts.

3. Special purposes and unique data problems require a regional accounts system to be conceived in its own particular context, even though there are some parallels that can usefully be drawn between national and regional accounts. Some of the differences between national and regional accounting that seem likely to emerge are:

a) Regional accounts need to be developed through a two-level approach, with the federal statistical agencies providing certain of the basic data for all of the regions and with the individual regions building on these and providing extensions and greater detail where needed for their own special purposes.

b) Unlike the national income and product and certain of the other national accounts, the key regional accounts may, for some period of time, be based on data provided periodically but not annually.

c) Because of this nonannual feature and because of the costliness of providing certain of the data needed for regional analysis, it seems likely that regional accounts, to achieve needed comprehensiveness, will have to depend on sample studies to a much greater extent than has been the case with national accounts.

There is a natural tendency to think of accounts in terms of the long experience with national accounts, so it is useful to focus on such differences to bring into relief the special problems posed by regional analysis and regional policy-making.

4. The most hopeful possibility at the present point in time, it seems to me, is to build a system of regional accounts centering on a geographic breakdown of the national interindustry sales and purchases estimates. First and foremost, the purchases-sales estimates will be integrated—both conceptually and statistically—with the national income and product accounts, so that much of the key data will be available on an industry basis. And, as the analysis in this paper suggests, the industry breakdowns are the critical ones for regional analysis and projection. Moreover, since the interindustry estimates will be built on the data provided by the economic censuses, a high proportion of the information is subject to regional identification.

Difficult conceptual and statistical problems are involved in such a geographic breakdown, to be sure, but they do not seem beyond solution. Whether Congress will be willing to provide the necessary funds is another matter, but one can be hopeful even on this score—particularly considering the history of interindustry work in federal agencies and the other statistical progress that has been made to date.

Even without nationally provided estimates, integrated regional interindustry input-output-income-product estimates might be prepared periodically within individual states and metropolitan communities that are interested in having such accounts as an aid to decision-making. Local data collection by governmental and private agencies could be organized through a co-ordinated effort geared to an accounts framework, and gaps in information could be filled through industry and household questionnaires. The large problem here (aside from the question of cost) is that of comparability, an essential in regional analysis. Thus, if there is to be no central provision of regional data (aside from those already being provided), a very serious effort would have to be made to get as much comparability in regional data collection as can be achieved without sacrificing special local data needs.

The provision of needed information within an accounts framework at the *local* level would of course be greatly simplified by a geographic breakdown of the national interindustry sales and purchases estimates: not only would interregional comparability be greatly enhanced, but the local efforts could be devoted to achieving greater refinement and detail.

5. A geographic breakdown of national interindustry estimates ideally would provide data for the states and the major SMSA's on (a) value added by each of the industry groups, (b) payments to labor and capital by each of these groups, (c) interindustry purchases and sales, including both intermediate and final users, (d) payments of taxes to the various governments and receipt of revenue by the various governments, and (e) "final demand" purchases of households and governments and as related to capital formation and exports.

The geographic breakdown of the national purchases and sales data will not provide information on an important item in regional analysis—the areas from which the purchases are made and to which the sales are made. This information, if it is to be made available, would have to come by way of industry questionnaires. It is the type of data that might be provided within particular regions, employing the national accounts data as a framework and as a basis for comparison.

6. Even taking the optimistic view that all the practical problems could be overcome and these "core" accounts established, quite a bit of information needed for regional analysis and projection would still be lacking. Some of this information could conceivably be provided through an extension of the interindustry accounts, while some would call for data outside the direct "core" accounts framework. Among the former are the manpower data which are needed for analysis of relative regional advantage, for localization analysis, and for impact analysis. We would want to have breakdowns for various characteristics of the labor force (sex, age groupings, race), for skills (i.e., skilled, semiskilled and unskilled), and for major occupational groupings. The manpower item in the interindustry accounts could be expanded by a supplementary table providing this information, estimating the average wage for each of the major categories rather than simply for all of labor within each industry

group. The availability and price of different types of labor are of prime importance in judging the relative attractiveness of different regions for industry.

To the greatest extent possible, the manpower items should be related to the household item in the final demand column, again through a supplementary table. Thus, for example, such interrelated information should suggest the number of nonwhite workers and households that might be expected from a given increase in industrial production within a metropolitan area. Also an anticipated increase in manpower could be related to rather specific types of household purchases, as well as to the increase in regional purchasing power generally. The main point is that bringing such information directly into the basic accounts framework offers the various advantages spelled out in Item 1 above and serves to strengthen the capacity for regional analysis and projection.

7. Some of the needed information does not readily lend itself to the basic accounts framework. This is especially true of the stock, or wealth, data where the double-entry feature is not particularly meaningful; that is, where either the asset value figure is of no great moment or there is no liability figure in a balance sheet sense—say with regard to public facilities financed by taxes. The major need is for data on stocks which are expressed largely in physical terms, as in the case of housing, land use, and productive capacity.

At the same time, there are important gains to be had by relating stock data to flow data wherever the relationship is analytically important and the value figures can somehow be produced. This is true, for example, in arriving at and analyzing gross and net investment figures which call for information on the depreciation and retirement of plant and equipment. These, in turn, call for estimates of depreciable assets. There is a great deal about regional economic growth and intraregional localization that can only be understood by relating current economic processes and flows to data on assets and liabilities and the physical units behind them. This can be exemplified by data on capacity, including both private productive capacity and public facilities. Estimates of regional growth must take into consideration excess productive capacity which can be employed to meet increases in demand: estimates of construction require production-capacity ratios as well as information on changing use of housing stock; estimates of needed public facilities call for information on capacity as well as on use of existing facilities. Where there are such analytically significant ties with the interindustry core accounts, there may be some value in placing such data into what might be called an "accounts-associated" category. The value would arise from the pressures to achieve account qualities of comprehensiveness, analytically meaningful categories, and periodicity. However, this point need not be stressed; the focus must be on a sense of priorities with regard to needed data, whether they are in accounts form, accounts-associated form, or just available. The main point is that certain stock data are important to regional analysis and projection.

The development of a regularized flow of information on stocks or wealth will obviously call for substantial research and experimentation far beyond anything done to date. We have the pioneering work of Raymond W. Goldsmith and others to build on, although it is only too evident that the problems of developing estimates at the regional level are particularly severe. On the other side, however, the need for stock data is more direct at the regional level—since so many of the decisions center on construction, redevelopment, and control of land use—so that great interest and practical use can be expected to follow on the heels of any progress made in this area.

8. Because so many decisions at the subnational level center on what we have called "localization," stock or wealth data should be organized in terms of *intra*regional subareas wherever possible. It is not enough to have figures for the inventory of housing within a given metropolitan area as a whole; it is important to know where the housing is located. The same is true for private productive facilities and government facilities, as well as for the major communication-flow channels such as highways, railways, and waterways. In general, the development of an accounts framework

for data needed in decisions focusing on intrametropolitan localization patterns deserves high priority in future research on regional accounts.

## COMMENT

*Jesse Burkhead*

*Syracuse University*

There can be no disagreement with the minimum position that Perloff has established in his excellent paper—that we will be better off in our ability to make public and private decisions that have an impact on economic growth if, a decade hence, we have developed the informational framework that he proposes. It may, however, be appropriate to ask what happens during the time that the prescription is being filled, and also to inquire about the state of affairs when we arrive at the Perloff prescription point.

While the prescription is being filled we are confronted with two conditions. The first is that public and private decisions affecting regional growth are now made, and will continue to be made, with inadequate utilization of the data already at hand. The spirit of development is so strong in some communities that industries are subsidized to enter in circumstances where neither public nor private officials stop to estimate the social costs of development and the spillovers from one jurisdiction to another. Within a region, an optimum development may be defined quite narrowly in terms of an increase in assessed valuation in one jurisdiction, or an increase in the volume of activity for banks and department stores. Success is measured by the degree to which the costs of development are shifted to other public jurisdictions and to unbenefited households and business firms. We could do much better than this. A rough but ready kind of benefit-cost analysis, with adequate attention to spatial distribution of benefits and costs, is possible for most communities within limits of existing data and techniques.

There is a second condition that affects both public and private decision-making at the regional level now, and that will continue for the next several years. This is the absence of considerable data, some expensive to secure and some inexpensive, which are conceptually rudimentary and are easily communicable to decision-makers. These include population estimates by municipalities and counties for intercensal years, annual estimates of county personal income, assessment and market value data by type of property for taxing jurisdictions within regions, and patterns of retail sales. Data of this kind, prepared on a systematic basis, could make major contributions to planning for urban renewal, to land-use planning in municipalities and counties, and to such mundane but important tasks as school facilities planning. In the development of comprehensive regional accounts, important as this is, we should surely not neglect the data gathering and policy problems to which economists can make an almost immediate contribution.

After we have arrived at the point when the Perloff prescription has been filled, we will find that there is no regional public or private authority with area-wide competence for decision-making. Indeed, there is no region that may be commonly defined for a wide range of problems. Land-use planning, recreation, transportation, water and air pollution, to name a few, may have no geographic boundaries in common, and definitions that reflect their changing character are necessary.

The lack of common geographic boundaries means that there are no common governmental boundaries for functional or program areas. In metropolitan regions in the United States the pervasive governmental characteristic is that of fragmentation. This multiplicity of governmental units is the product of historical accident, the inadequacies of local government tax bases, and an unbelievably complex pattern of behavior based on free choice, ethnic prejudice, and income-class motivations. No public agency does now or can be expected to speak for this complex and to make general government decisions on a regional basis.

The absence of central administrative responsibility for regional decision-making in the public sector implies an absence of central responsibility for the economic welfare of the region. That which contributes to the welfare of one region does not necessarily contribute to the welfare of other regions. Similarly, that which contributes to the welfare of the region as a whole does not contribute uniformly to the welfare of the subunits of the region.

The nonadditive nature of regional welfares is everywhere apparent. Moreover, within regions, each governmental jurisdiction has the ability, limited or extensive as the case may be, to compete with other jurisdictions in that region for a greater than average share of regional growth. Alternatively, some jurisdictions, motivated by a mixture of aesthetic considerations and exclusionist sentiment, may be able to resist an economic growth that would otherwise occur. We do not have a regional social diseconomies board with authority to parcel out the spillover gains and losses to effect a rational allocation of resources in accordance with relative improvements in welfare among subunits of the region. In fact, in the typical metropolitan region, with its mixture of communities that compete eagerly for economic growth and those that as firmly resist it, the very concept of a rational regional growth pattern is most elusive. Regional isolationism and regional fragmentation will not contribute to a rational allocation of either public or private resources. Regional accounts, with their implicit emphasis on a subnational approach to welfare measurements, must be used with caution if they are not to encourage a parochialism in policy-making that is already far too dominant in many communities and areas.

3

# INVESTMENT RETURN AS A MEASURE OF COMPARATIVE REGIONAL ECONOMIC ADVANTAGE

*George H. Borts and Jerome L. Stein*[1]

*Brown University*

PREDICTION OF relative growth rates is a major concern of research in the regional and urban field. Economists are frequently called upon to indicate the pattern of industrial growth in individual areas as well as the differential growth pattern among geographic areas. Both problems have immediate interest for the formulation of local policy. If industry is to be stimulated, it is fruitful to stimulate the growth of those industries that have the greatest chance of survival once the temporary shelter of government help is withdrawn. In addition, the magnitude of support a community receives is frequently determined by an estimate of the way a region will fare in the absence of such support.

In this paper we develop a predictive method which is derived from the theory of marginal productivity. This theory is applicable as a predictor both of relative growth differences among regions

[1] The authors are indebted to Harold J. Barnett, Edward B. Berman, Daniel Creamer, Richard Muth, Harvey S. Perloff, and Sidney Sonenblum for helpful comments on earlier drafts of this paper.

and of industrial growth patterns within a region. We concentrate on the latter, but in later sections show how the theory can be applied to other aspects of regional analysis. In the last section of the paper we present our conclusions concerning the rationale of collecting data for regional accounts.

### *The Use of Ad Hoc Methods vs. the Use of Price Theory to Predict Growth Rates*

The theory of the competitive industry is a logical development of the assumption that, given a set of alternative actions, the business firm will select the action which it expects to be most profitable. This theory has been a fruitful source of hypotheses concerning the behavior of the economy; and we have seen the success of economic analysis in explaining the complex world of our experience.[2]

As Richard Muth said at this Conference: "All too often economists in tackling a new problem promptly forget their economic theory and resort to vague guesses, naive extrapolations of past behavior, or theories quite unrelated to traditional analysis in search of a solution. This tendency is especially evident in the area . . . [of regional economic growth]. Having seen the success of traditional economic analysis in handling a wide variety of problems, I would tend to place more confidence in its predictions in new areas than in those of *ad hoc* methods."

We demonstrate that our method of prediction based upon economic theory is *at least as good* a predictor of growth as *ad hoc,* or mechanical, methods based on rational behavior.[3] Such methods suffer from three deficiencies. First, there is clearly no reason to expect that the same set of *ad hoc* methods will work for a subsequent period. Second, there is no understanding of why these *ad hoc* methods were chosen. Third, an *ad hoc* method can hardly serve as a guide to public policy: it cannot suggest the factors

[2] See J. L. Stein, "The Predictive Accuracy of the Marginal Productivity Theory of Wages," *The Review of Economic Studies,* 25, No. 3 (1958).

[3] By rational behavior we mean the behavior which will lead to profit maximization.

which were responsible for the high or low growth rates; it cannot suggest what policies will or will not be efficacious in raising the region's rate of growth.

## *I. AD HOC METHODS OF PROJECTION*

In the past, investigators have predicted regional growth patterns through a variety of methods, most of which involve a weighting and projection of current or past rates of change. While these procedures are justifiable as exploratory devices, they have failed to provide accurate forecasts; neither have they offered answers to questions of causation. Two such procedures have been frequently used, and they richly deserve retirement.

First is the application of past trends to predict current or future changes among regions or among the industries of a given region. When public officials ask, "Which industries will grow most rapidly in this locality?" the answer this technique provides is: "The industries that have already grown most rapidly in this locality or in neighboring localities." We shall see later the errors such a method can yield.

Second is the weighting of growth trends. This method assumes that the growth rate of a region is determined by the composition of the industries it contains. If these industries grow at a particular rate nationally, it is assumed that they will grow at the same rate in the region under examination. In order to know how fast a region is to grow, one first predicts the national growth rate of the region's industrial components. The predicted growth rates are then weighted by the proportion in which the components are presently combined in the region; this yields the projected growth rate for the region.

As a test of the usefulness of this second method, we have constructed projected growth rates of manufacturing employment for the forty-eight contiguous states. The projections were built up under highly favorable assumptions. Rather than attempt to predict the growth of the national industrial sectors, we used their actual growth rate over the period for which the projection was made.

Thus in projecting state growth between cycle peaks, we used the actual growth rates of twenty national industry groups.[4] The projected growth rate of manufacturing employment is then the rate the state would experience if each of its industrial components grew at the national rate.[5]

The computation of projected and actual growth rates in manufacturing employment was carried out for the forty-eight states over the four time intervals: 1919-29, 1929-47, 1948-53, and 1948-57.

It will be noted that the terminal dates of all four intervals correspond roughly to peak business cycle dates. By measuring changes between similar phases of the business cycle, cyclical influences are largely eliminated. For these trends are consistent with any value of cyclical amplitude.

The actual and projected growth rates are compared through the use of the Spearman rank correlation coefficient. These values are shown on page 73.

[4] The twenty groups correspond roughly to the following industries from the Standard Industrial Classification used in the *Census of Manufactures* for 1947.

| Durable Goods | | Nondurable Goods | |
|---|---|---|---|
| *Group No.* | | *Group No.* | |
| 24 | Lumber products | 20 | Food and kindred products |
| 25 | Furniture and Fixtures | 21 | Tobacco manufactures |
| 32 | Stone, Clay and Glass products | 22 | Textile mill products |
| 33 | Primary Metal industries | 23 | Apparel and related products |
| 34 | Fabricated Metal products | 26 | Paper and allied products |
| 35 | Machinery (except electrical) | 27 | Printing and publishing |
| 36 | Electrical Machinery | 28 | Chemicals and allied products |
| 37 | Transportation Equipment | 29 | Petroleum and coal products |
| 38 | Instruments | 30 | Rubber products |
| | | 31 | Leather products |
| | | 391, 396 | Jewelry and Silverware |

[5] Let $L_1 = \frac{l_{11}}{l_{10}}$ where $l_{11}$ is the employment in U. S. industry 1 in year 1, and $l_{10}$ is employment in the same industry in the base year 0. Let $W_{io}$ be the proportion of a state's employment in industry 1 in year 0. Then the projected employment growth rate is $\Sigma_i L_i W_{io}$.

RANK CORRELATION BETWEEN PROJECTED AND ACTUAL GROWTH RATE OF MANUFACTURING EMPLOYMENT, 48 STATES, IN FOUR TIME INTERVALS

| | |
|---|---|
| 1919–29 | − .0334 |
| 1929–47 | + .0601 |
| 1948–53 | + .4348* |
| 1948–57 | + .1494 |

* significant at 5%

It can be seen that only one of the correlations is significantly different from zero. In three of the four periods examined, the state industrial components grow at rates which bear little relation to their national counterparts. In the period from 1948 to 1953, we have evidence from earlier work that demand changes dominated the interstate growth pattern.[6]

The results of these comparisons show the weakness of using weighted extrapolations of national growth patterns to predict the growth rate of a particular region. It is clear from the above correlations that the growth rate of a region does not depend upon the national pattern of demand facing its industries. Differential growth rates are produced because a given industry grows at different rates in different regions.

## *II. PREDICTION BASED ON ECONOMIC THEORY*

We derive and apply a method of predicting regional and industrial growth which is based on the theory of the competitive industry. We also show the extent to which these methods are limited in their scope by the available data on income and wealth categories for different regions.

[6] See G. H. Borts, "The Equalization of Returns and Regional Economic Growth," *American Economic Review*, 50 (June 1960).

The 1948-53 period is characterized by a number of conditions which support this view:

a) High-wage industries enjoyed the greatest relative price increases and wage increases, and

b) Rates of growth of nonagricultural employment and capital were observed to be higher in high-wage states than in low-wage states.

### *The General Rationale of Our Predictive Method*

An estimate of the relative marginal rate of return on investment between two regions is used to predict which region will have the greater growth of employment. The economic logic of this view is that businessmen strive to select the alternative which is expected to be most profitable. They will invest in, and thus expand, those industries where the marginal return on additional capital invested is expected to yield the greatest return per dollar of investment. If firms act to maximize profits, as we assume, then an above average rate of expansion will take place in the areas where the marginal rate of return on investment is expected to be higher than the average. Insofar as capital and labor are complementary in production, the expansion will be reflected in a growth of employment. A high correlation will be observed between the growth of output and the growth of employment, if the industry expands via an increase in the number of plants.

Although little is known concerning the formation of expectations, some sensible statements can be made on the subject.

"It would be foolish, in forming our expectations, to attach great weight to matters which are very uncertain. It is reasonable, therefore, to be guided to a considerable degree by the facts about which we feel somewhat confident, even though they may be less decisively relevant to the issue than other facts about which our knowledge is vague and scanty. For this reason the facts of the existing situation enter, in a sense disproportionately, into the formation of our long-term expectations; our usual practice being to take the existing situation and to project into the future, modified only to the extent that we have more or less definite reasons for expecting a change." [7]

The businessman desires to make an investment in a given industry by, say, establishing a new plant. His choice of location of the investment depends upon factors of past experience and future expectations. Future expectations are largely based upon past experience. Consequently, major factors which influence the loca-

[7] J. M. Keynes, *The General Theory of Employment, Interest and Money* (New York: Harcourt, Brace and Co., 1936), p. 148.

tion of this investment are the marginal rates of return that have occurred in the various possible geographical areas in which investments have been made. His choice will be that area with the highest marginal rate of return in the recent past. Experience will indicate whether the businessman was correct in his expectation of future marginal rates of return. If, *ex post*, the marginal rate of return on investment is lower in area $X$ than in area $Y$, expectations will be revised in the light of this information. Future investment will be directed toward area $Y$ rather than toward area $X$. Expectations are constantly revised on the basis of experience. But investment in any given period will be guided by the marginal rates of return that materialized in the previous period.

Schematically, our theory is this:

relative marginal rate of return on investment in $(t-1)$ $\longrightarrow$ relative rate of growth of employment in $t$.

The logic underlying this predictive device is an outcome of the theory of the competitive industry. Under static conditions, each firm will be earning a rate of return such that there is no net addition to the capital employed in the industry. That is, the schedule of marginal efficiency of investment as a function of the interest rate intersects the normal rate of return at a zero rate of net investment. Let us split the industry into two sectors which are for the moment distinguished only by their geographic location: one sector in New England, and one in the rest of the country. Let us also assume that neither sector is in long-run equilibrium, but is instead adjusting to past changes in demand, costs, and technology. For each sector we observe a point on its marginal efficiency of investment schedule. The co-ordinates of the point are the observed value of the marginal efficiency and the percentage growth of capital in the sector. These are illustrated in Figure 1. The percentage growth of capital, rather than the rate of investment, is used because of differences in the absolute size of sectors. As a practical matter we are able to identify one schedule as higher than the other because we observe the marginal efficiency to be of greater value in one sector. That is, if investment is greater in the

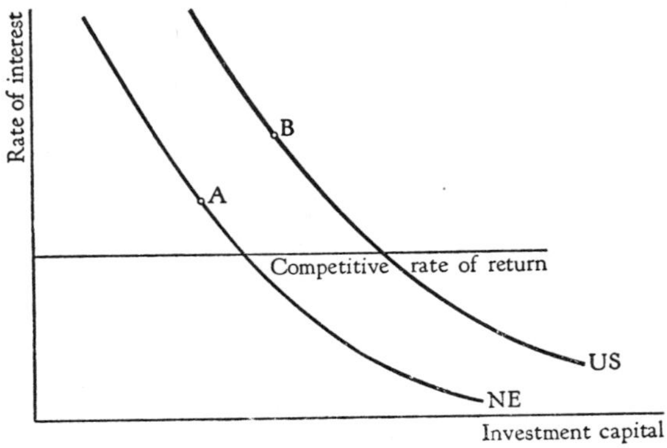


*Figure 1.*

United States, we will observe a greater marginal efficiency for the United States. This is shown in the diagram. We would observe point *B* for the United States and point *A* for New England. This arises from our method of computing the marginal efficiency. If we could truly observe the profit results of incremental investment within each firm, then the computed values of the marginal efficiency would equal the same interest rate and the schedules would differ only by their respective rates of investment. However, we can only observe the entire investment experience of the firms in the sector over a given period of time. What we compute as the marginal efficiency is actually the average rate of return on the investment which occurred. This will be greater in the sector which is undergoing the greater expansion. This behavior is consistent with an expansion in the number of firms or the growth in the scale of existing firms. In each case there is a greater demand for labor.

### *Reasons for Differences in Marginal Rates of Return on Investment*

Why does a higher marginal efficiency in Sector *A* imply a more rapid growth of employment in Sector *A*? The answer in simplest terms is that an increase in the number of firms in a sector implies an increase (shift) in the sector's demand schedule for labor.

Finally we must ask, under what conditions will one sector have a higher schedule of marginal efficiency of investment than the other? To what extent must the sectors differ in order to generate such behavior?

1. One possible answer must be rejected out of hand because of our fundamental assumption that the two sectors belong to the same national industry. That is, if the two sectors are really facing different demand schedules, then they cannot both be selling in the same national market.

2. A second answer is that differences in technology in the two sectors are responsible for differences in the marginal efficiencies. In the strictest sense of the concept, this is hardly possible in manufacturing, especially where we are examining the behavior of regional sectors in the same national economy. Manufacturing technology, in the sense of the state of the arts and sciences, is in fact perfectly diffused among U. S. regions. The differential willingness of entrepreneurship to adopt technological change should not properly be considered under this classification.

3. Changes in transport costs might be responsible for changes in relative marginal efficiencies. For example, a rise in transport costs on materials would reduce the marginal efficiency of investment for firms most distant from sources of supply.

4. The most likely explanation is that differences in the supply conditions of resources other than capital are responsible for differences in the marginal efficiencies and in growth rates.[8] This will occur when there is an expansion in the demand for the industry's product. Let us assume that this increased demand affects both sectors in equal measure. However, in one sector the elasticity of supply of labor is greater than in the other. The sector with

[8] The theory developed here is based upon Jerome L. Stein, "A Theory of Interstate Differences in the Rates of Growth of Manufacturing Employment in a Free Market Area," *International Economic Review*, 1, No. 2 (May 1960).

greater supply elasticity will experience greater growth of employment, and in the process a greater growth of capital. While this process of adjustment is going on and before it has been completed, the expanding sector will also evidence a higher return on new investment per firm than the other sector.

An exact specification of the conditions under which this will occur are:

a) Perfect competition or the same degree of monopoly in both sectors.
b) The f.o.b. price of the product is the same in both sectors.
c) Each firm in the industry faces the same production function which is homogeneous of degree unity, but there is an external limit to the size of each firm.
d) Each firm in the industry faces the same price of capital.
e) The elasticity of labor supply is greater in Sector II than Sector I.

Assume the industry is initially in long-run equilibrium such that each sector pays the same wage and yields the same return on capital. For graphic purposes of exposition, assume that each sector initially hires the same quantity of labor. That is, this equilibrium happens to occur at an intersection of the two labor supply functions.[9] This is illustrated in Figure 2. Now assume that there

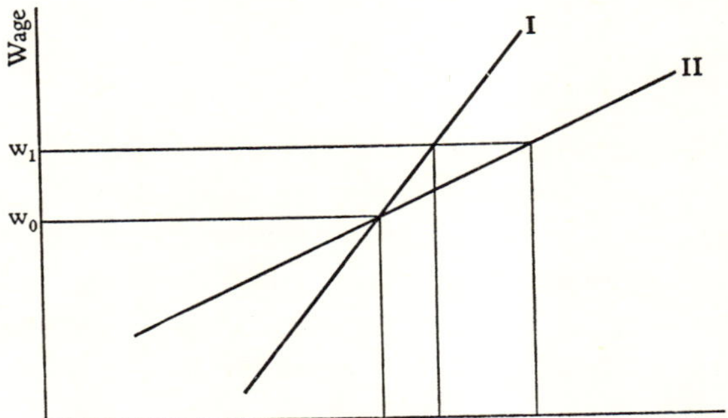


*Figure 2.*

[9] This is just a simplifying assumption for purposes of exposition.

is an increase in the demand for the products of each sector. In the final equilibrium, the wage must again be the same in each sector, and a greater expansion of employment and capital will have occurred in Sector II. Suppose, however, that the expansion is completed in Sector I, but has not yet terminated in Sector II, so that employment in Sector II is also at $L_1$. This will yield a higher return to capital in Sector II as a result of the lower wages, and it will also imply in subsequent periods an expansion of employment in Sector II which exceeds that in Sector I.

Thus, our theory leads us to expect that higher rates of return on investment will be accompanied by higher rates of growth of employment. Note that this conclusion is consistent with a dynamic growth process in each sector such that neither sector ever reaches a final equilibrium where investment would fall to zero.

### *A Measure of the Marginal Rate of Return on Investment*

Ideally, we seek to calculate the marginal rate of return in new investment by comparing the increment of a flow of proprietary income to the purchase price of additional assets. The marginal efficiency of investment is then the discount rate $r$ which equates the discounted stream of returns with the purchase price of the additional assets $P$. The return $R$ is total revenue minus total operating costs and taxes. In the continuous case

$$(1)\ P = \int_{t=0}^{n} R(t)e^{-nt} = \bar{R}\int_{t=0}^{n} e^{-nt}, \text{ where } n \text{ is the life of the asset.}$$

The first equality is the standard definition of the marginal rate of return, $r$. The second equality states that there exists a return $\bar{R}$ which can be considered a constant.[10]

If we treat all investments as perpetual, i.e., we assume that replacement is always equal to the current period depreciation,

[10] This is the mean value theorem for integrals.

$\int^{\infty} e^{-nt} = \frac{1}{r}$, and $\bar{R}$ is now net of depreciation expense. Hence

$$(2) \quad P = \bar{R}\left(\frac{1}{r}\right),$$ the well-known capitalization formula.

We would like to measure $(\bar{R}/P) = r$, for each industry and region. However, numerous compromises with the theory must be made in view of the available data.[11]

1. The variable $P$ consists of an increment of assets, regardless of whether it is in fixed or working capital.

There are no data on working capital by industry and region. We are, therefore, forced to use the marginal rate of return on additions to plant and equipment alone. Symbolically it means that we can only observe some fraction of $P$. We would like to calculate $\bar{R}/P$, but we end up calculating $\bar{R}/I$, where $I$ is investment in plant and equipment. $P - I$ consists of investment in working capital.

Insofar as the two regions compared have the same ratio of $I$ to $P$, no difficulty arises in comparing their marginal rates of return. We would expect that, for a given industry, interregional variations in $I/P$ are small; and hence our compromise with the data is not significant. Our independent variable is always the *relative* marginal rates of return in a given industry, among regions. For this reason, the omission of data on working capital does not seem to be serious.

2. It is impossible to know in advance the future income stream of an investment. Furthermore, the limitations of the data are such that we can measure the value of income to property for only a discrete number of periods. Consequently, we measure the flow of new proprietary income arising from investment as the difference between the level of proprietary income in two years.

[11] This section has benefited from the perceptive comments of Daniel Creamer, who discussed this paper at the Conference.

The particular years chosen are dictated by the availability of data. Because of our concern with manufactures, we have used the dates of the last three Censuses of Manufactures for the United States, 1939, 1947, and 1954.

Over the $n$ intercensus years proprietary income rises by $R_1'n$ in region one and by $R_2'n$ in region two. On the average, annual proprietary income rises by $R_1'$ and $R_2'$ in the two regions.

Our *measure* of the marginal efficiency of investment $m_1$ is $R_1'/I_1$; and of $m_2$ it is $R_2'/I_2$. The *true* $r_1$ is equal to $\bar{R}_1/P_1$ and the true $r_2$ is $\bar{R}_2/P_2$.

If $R_1'/\bar{R}_1$ and $R_2'/\bar{R}_2$ are not very different,[12] then $m_1/m_2$ is approximately equal to $r_1/r_2$. That is, no significant bias would be introduced into our measures of relative rates of return by choosing these intercensus years as our sources of data on changes in proprietary income.

3. When computing the marginal efficiency of investment for a 4-digit industry as identified in the Census of Manufactures, the following definition is employed:

| | | |
|---|---|---|
| | (1) | Total value of shipments |
| less | (2) | Cost of materials, supplies, containers, fuels, purchased electric energy and contract work |
| equals | (3) | Value added |
| less | (4) | Wages and salaries |
| equals | (5) | Gross proprietary income. |

Gross proprietary income includes depreciation, inventory decumulation, profits, interest, taxes, rents, insurance premiums, legal fees, and other business services purchased. Ideally, to be applicable to regional analysis, the figure should be reduced to net profits before the deduction of federal income taxes, interest, and rents. Thus, it would be independent of the type of financing used and of the past decisions to own or rent plant and equipment. Further refinements of the data could conceivably be made if we had tax returns on a plant and region basis.

[12] And if $I_1/P_1$ is approximately equal to $I_2/P_2$, where the subscripts refer to regions.

4. The marginal return on net investment in plant and equipment cannot be calculated for industries identified in the Census of Manufactures. Net investment is not reported in the Census. What is reported is expenditures for plant and equipment, which is essentially a gross investment concept. What we have computed for these categories is the marginal rate of return on gross investment. This measure is then

$$m = \frac{\textit{change in Gross Proprietary Income}}{\textit{Expenditures for Plant and Equipment}}$$

In our other studies the marginal rate of return on gross investment has been adjusted so that it may reflect the return on net investment. This is frequently necessary in an analysis of the regional aspects of growth. Gross investment may include two categories which are not properly part of net investment. One is the replacement of obsolete equipment; the other is the purchase of old plant and equipment which is temporarily undervalued because of a secular decline in the industries using the equipment. In either case, a high rate of return on gross investment may not be indicative of long-run growth potential in a particular industry or area. This problem arose in other studies where the authors attempted to specify the particular New England industries in the chemical and metal working field which had high marginal rates of return on net investment.[13]

[13] Jerome L. Stein, *Economic Factors in the Location of Industry, Part One: The Chemical Industries* [1957]; George H. Borts, *Economic Factors in the Location of Industry, Part Two: The Metal Working Industries* [1960] (Providence, R. I.: Brown University, College-Community Research Program).

In these studies the problem was handled by constructing estimates of depreciation on a regional and industrial basis. We used the ratio of gross investment to value added as a rough measure of the rate of net investment for an industry group in a region.

Write this as $\alpha_{ij} = \dfrac{\text{Gross Investment } ij}{\text{Value Added } ij}$

where $i$ stands for a 4-digit industry group and
$j$ stands for the region.

We then say that if

$m_{ij} > m_{iUS}$ (marginal rate of return on gross investment)

$\alpha_{ij} > \alpha_{iUS}$ (ratio of gross investment to value added),

then the industry $i$ has a higher rate of return on net investment in the $j^{\text{th}}$ region than it has in the United States as a whole.

The marginal rate of return on gross investment, $m$, rather than on net investment, is used as the independent variable in this study. Numerous compromises with the theory have been made, in order to develop a predictive tool. The residuals between the actual and predicted growth rates are the sole indicators of how significant are the compromises made. If the predictive accuracy of our theory is high, then the assumptions made (concerning $I/P$, $R'/\bar{R}$, etc.) are not serious oversimplifications of reality. The crucial question is: how good a predictor do we have in view of the data limitations?

*Predictions*[14]

1. We are interested in predicting the growth of employment in specific industries from a knowledge of the rate of return on investment in those industries. We calculate this rate of return for both prior and concurrent periods, relying mainly on the former as our predictor. The industries are the 4-digit members of Census Group No. 35, Non-Electric Machinery. We predict the growth for the New England region; but this method could be used for any region or for any industry for which data are available.

Define

$$\mu_i = m_{NE} - m_{US}.$$

This is the difference between the return on gross investment in the $i^{th}$ industry in New England and the return for that industry in the United States.

Also define

$L_{ij}$ = the rate of growth of employment in the $i^{th}$ industry and $j^{th}$ region, between two periods.

---

The use of the ratio $\alpha$ embodies the assumption that the ratio of depreciation to value added is roughly the same for all geographic sectors of a 4-digit industry. When this assumption is satisfied, $\alpha$ expresses regional differences in the ratio of net investment to value added. This assumption leaves much to be desired, for the higher the rate of expansion of reproducible capital, the lower is the ratio of depreciation to gross investment. Since no data on depreciation are available, we are forced to use gross investment as our denominator.

[14] This section has benefited greatly from the comments made by Richard Muth on the earlier draft of this paper presented to the Conference.

Define

$$\lambda_i = L_{\widehat{iNE}} - L_{\widehat{iUS}}.$$

This is the algebraic difference between the rate of growth of employment in the $i^{th}$ industry in New England, and the rate for the same industry in the United States.

The values of the $m_i$, $L_i$, $\mu_i$, and $\lambda_i$ are shown in Tables 1 and 2. They have been computed for the intervals 1939-47 and 1947-54. These periods were chosen because of the availability of data from the Census of Manufactures. The industry groups are those 4-digit members of Group 35 which are represented in New England for the three dates.[15]

Correlations between $\lambda$ and $\mu$ are significant at the 5% level. For the period 1939-47, the following regression relation was obtained:

$$\lambda = \underset{(11.1254)}{30.1752\mu} - 14.6925, \qquad r = .5496,$$

[15] In a number of cases it was necessary to reconstruct the composition of the 4-digit group in New England for 1939. The reason is that data on gross investment, wages and salaries, and value added are not available on a region basis for 1939. Further, gross investment was not available on a region or state basis in 1939.

Estimates of wages and salaries and value added for 4-digit industries in New England were derived as follows:

1) In a given 4-digit industry, we know from the Census the production worker employment in 1939 for New England.

2) We also know for each New England state the number of production workers, the wage bill, and the value added in the industry. The information given for the states will fall short of the New England total, because the disclosure rule will eliminate some of the state information.

3) The information for the states was blown up to the New England total by using the ratio of total production workers in New England, to the production workers reported by state. This involves using the assumption that the wage bill per production worker and value added per production worker are the same in the known and unknown sectors of the industry. For the 19 industries data available for 1939 included on the average 75% of the total production worker employment in each of the categories.

Estimates of gross investment for 1939 for New England 4-digit industries were derived as follows: For each industry in New England, the ratio of gross investment to value added was formed for 1947. This ratio was then applied to the 1939 value added for each industry. This assumes that the ratio remains the same over time, but of course, differs by industry and region.

TABLE 1. VALUES OF MARGINAL EFFICIENCY OF GROSS INVESTMENT, $m$, GROWTH OF EMPLOYMENT, $L$, FOR NEW ENGLAND AND UNITED STATES, 1939-47, BY 4-DIGIT INDUSTRY IN THE MACHINERY GROUP

| Industry No. | Title | 1<br>$m_{NE}$ | 2<br>$m_{US}$ | 3<br>$\mu = 1-2$ | 4<br>$L_{NE}$ | 5<br>$L_{US}$ | 6<br>$\lambda = 4-5$ |
|---|---|---|---|---|---|---|---|
| 3541 | Machine Tools | .0686 | .8601 | − .7915 | 133.89 | 148.37 | − 14.48 |
| 42 | Metalworking Machy. nec | 1.4239 | 2.1053 | − .6814 | 191.62 | 240.40 | − 48.78 |
| 43 | Cutting Tools, Jigs, Fixtures | 1.3184 | 1.4548 | − .1364 | 249.28 | 288.46 | − 39.18 |
| 51 | Food Products Machy. | − .9724 | 1.2085 | −2.1809 | 92.70 | 203.65 | −110.95 |
| 52 | Textile Machy. | 1.1694 | 1.4633 | − .2939 | 206.80 | 201.03 | + 5.77 |
| 53 | Woodworking Machy. | 5.4631 | 2.1779 | +3.2852 | 363.29 | 345.93 | + 17.36 |
| 54 | Paper Industries Machy. | 0.9595 | 1.0165 | − .0570 | 235.33 | 240.54 | − 5.21 |
| 55 | Printing Trades Machy. | 2.0665 | 1.7159 | + .3506 | 276.67 | 204.77 | + 71.90 |
| 59 | Special Industry Machy. nec | 1.0743 | 1.4325 | − .3582 | 218.80 | 247.46 | − 28.66 |
| 61 | Pumps and Compressors | 3.0791 | 1.5397 | +1.5394 | 279.09 | 201.16 | + 77.93 |
| 62+63 | Elevators, Escalators, Conveyors | 4.1044 | 3.1842 | + .9202 | 188.74 | 263.82 | − 75.08 |
| 66 | Power Transmission Equip. | 0.8993 | 1.1793 | − .2800 | 156.25 | 241.58 | − 85.33 |
| 69 | Gen'l. Industry Machy. nec | 2.0889 | 1.6074 | + .4815 | 310.94 | 205.83 | +105.11 |
| 72 | Typewriters | 1.3513 | 0.9290 | + .4223 | 125.25 | 147.19 | − 21.94 |
| 83 | Sewing Machines | 0.6503 | 0.8782 | − .2279 | 165.69 | 161.31 | + 4.38 |
| 85 | Refrigeration Machy. | 0.7221 | 1.0228 | − .3007 | 212.32 | 274.68 | − 62.36 |
| 91 | Valves and Fittings | 0.3497 | 1.3062 | − .9565 | 136.21 | 204.78 | − 68.57 |
| 93 | Ball & Roller Bearings | 1.0098 | 0.7193 | + .2905 | 205.91 | 235.21 | − 29.30 |
| 99 | Machine Shops | 2.2195 | 1.2710 | + .9485 | 287.67 | 199.88 | + 87.79 |

*Source: Census of Manufactures, 1947 and 1939* (Washington: U. S. Department of Commerce).

TABLE 2. VALUES OF MARGINAL EFFICIENCY OF GROSS INVESTMENT, $m$, GROWTH OF EMPLOYMENT, $L$, FOR NEW ENGLAND AND THE UNITED STATES, 1947-54

| Industry No. | 1 | 2 | 3 | 4 | 5 | 6 |
|---|---|---|---|---|---|---|
| | $m_{NE}$ | $m_{US}$ | $\mu$ = 1–2 | $L_{NE}$ | $L_{US}$ | = 4–5 |
| 3541 | 1.5875 | 1.3683 | + .2192 | 117.08 | 115.65 | + 1.43 |
| 42 | 0.5115 | .8661 | − .3546 | 87.14 | 108.15 | − 21.01 |
| 43 | 0.4639 | .5602 | − .0963 | 105.28 | 137.57 | − 32.29 |
| 51 | 2.1815 | .0370 | +2.1445 | 318.43 | 89.92 | +229.51 |
| 52 | −.4581 | −.3302 | − .1279 | 54.66 | 68.30 | − 13.64 |
| 53 | 3.4703 | .2469 | +3.2234 | 115.92 | 73.76 | + 42.16 |
| 54 | .2758 | .3948 | −0.1190 | 89.18 | 87.80 | + 1.38 |
| 55 | .2364 | .2227 | + .0137 | 77.69 | 88.26 | − 10.57 |
| 59 | .8690 | .2366 | + .6324 | 89.69 | 98.25 | − 8.56 |
| 61 | 1.6217 | .9328 | + .6889 | 79.37 | 107.45 | − 28.08 |
| 62+63 | 1.4070 | 1.0772 | + .3298 | 133.67 | 122.47 | + 11.20 |
| 66 | .6665 | .5100 | + .1565 | 122.08 | 91.82 | + 30.26 |
| 69 | −.0003 | .0756 | − .0759 | 51.59 | 80.14 | − 38.55 |
| 72 | .0674 | .3411 | − .2737 | 68.13 | 69.29 | − 1.16 |
| 83 | .0963 | −.0052 | + .1015 | 93.39 | 81.11 | + 12.28 |
| 85 | .8811 | .4267 | + .4544 | 121.70 | 99.69 | + 22.01 |
| 91 | .7649 | .7285 | + .0364 | 103.57 | 94.11 | + 9.46 |
| 93 | −.0784 | .0879 | − .1663 | 80.15 | 95.30 | − 15.15 |
| 99 | 1.0698 | .9487 | + .1211 | 167.19 | 166.23 | + .96 |

*Source:* With the exception of groups 3572 and (3562 + 3563), the entries of this table are derived from Jerome L. Stein and Mark B. Schupack, *The New England Machinery Industry in 1970* (Federal Reserve Bank of Boston, 1960).

For the 1947-54 period, the following regression was obtained:

$$\lambda = \underset{(17.69)}{48.94}\,\mu - 6.68, \qquad r = .7892.$$

There is a good statistical reason why the regression coefficient and the correlation coefficient of the regression for the 1939-47 period are lower than they are in the regression for the 1947-54 period. This results from the greater errors of measuring gross investment in the earlier period.[16]

Since no data on gross investment for 1939 were available for New England 4-digit industries, investment was estimated by multiplying the ratio of gross investment to value added in 1947

[16] We are indebted to Richard Muth for this observation.

by the 1939 value added. The denominator of $m$ in the 1939-47 period is an average of the gross investment *estimate* in 1939 and the *reported* gross investment figure in 1947. On the other hand, the denominator of $m$ in the 1947-54 period is an average of two *reported* gross investment figures.

There is evidence that the ratio of gross investment to value added differed between 1939 and 1947. Consequently, the errors of measurement of the $m$ variable were probably greater in the earlier period. It can easily be shown that errors of measurement in the independent variable reduce the regression and correlation coefficients.[17] A comparison of the two regression equations above supports this hypothesis. If this interpretation is correct, then the relation between $\mu$ and $\lambda$ may have been more stable than is indicated by the above equations.

2. Our prediction of growth based on the marginal efficiency concept is the following. The growth position of an industry is determined by its relative rate of return in the prior period. The relation is given by the regression line for the prior period. In addition, where an industry's growth was underestimated in the prior period, its subsequent growth position will conform more closely to the regression. Finally, if the regression fails to predict the industry's growth for the later period, this is due to a change of

17

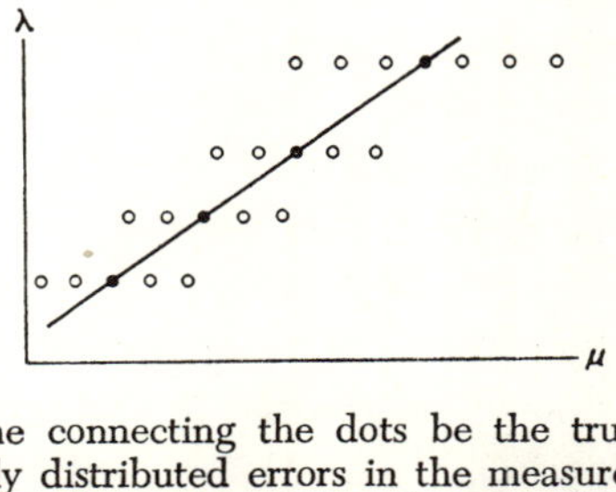


Let the straight line connecting the dots be the true regression equation. As a result of normally distributed errors in the measurement of $\mu$, our observations are the *o*'s. The dispersion of the *o*'s raises the *measured* variance of the $\mu$'s. On the other hand, it does not affect the covariance $\Sigma(\lambda - \bar{\lambda})(\mu - \bar{\mu})$, because for a given $\lambda$, the $\Sigma(\mu - \bar{\mu}) = 0$. Since the regression coefficient is cov $(\lambda, \mu)$/var $\mu$, the measured regression coefficient is less than the slope of the true line. A similar argument explains the lower correlation coefficient.

the marginal rate of return. These predictions can be visualized with the aid of Figure 3, showing the regression of $\lambda$ on $\mu$ for the

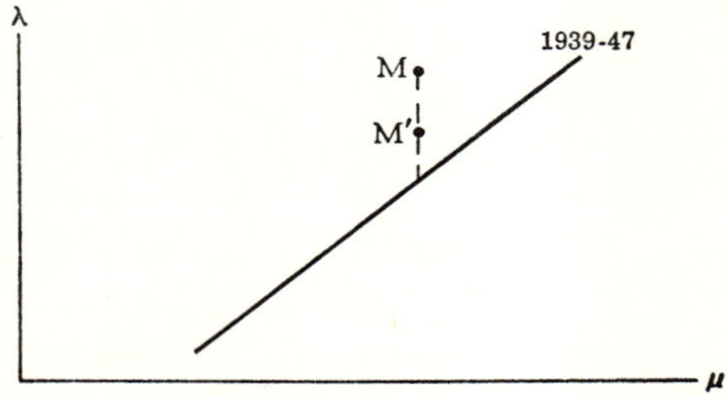


*Figure 3.*

period 1939-47. We shall use the regression line for 1939-49 to predict the growth of industry group *i* in 1947-54, despite the difficulties arising from errors of measurement. With better data for the initial period, we would expect better predictive accuracy. That is, we shall use the predicted $\lambda_i$ from the regression to be the $\lambda_i$ the industry will have in the subsequent period. Since $\lambda_i = L_{iNE} - L_{iUS}$, a projection of the U. S. growth rate in the industry will give enough additional information to provide an estimate of $L_{iNE}$ the New England growth rate. By doing this we are eliminating any error from the prediction of the U. S. growth rate in the industry and predicting regional growth as the deviation from the U. S. rate.

Let us denote the coordinates $(\lambda, \mu)$ for industry *M* in 1939-47 as the point *M* in the diagram. Similarly the point *M'* represents the actual value of $\lambda$ for the 1947-54 period. If $\lambda$ of *M* was underestimated in the first period, then in the second period $\lambda$ of *M'* must be closer to the regression line.

Finally, we conclude from this that deviations of $\lambda$ of *M'* from the predicted regression line are due to changes in the value of $\mu$ for the subsequent period.

Note there is nothing in our analysis which implies stability of the $\mu_i$. What we do assume is stability of the relation between $\lambda$ and $\mu$ so that it can be used for predictive purposes. Unfortunately, we could not derive this relationship from better data.

*Test of Predictive Power*

Tests of the predictive power of our technique were run against a number of *ad hoc* projection techniques, i.e., techniques that have no rationale but were devised for this specific situation. The *ad hoc* techniques were invented to serve as bases of comparison between our predictive method and all other contenders. Our motivation was: name any conceivable technique and we shall compare its accuracy with ours.

The *ad hoc* techniques took the following possible forms:

(A) Assume $\lambda_{i47} = \lambda_{i39}$ for each industry. This simply projects the $\lambda_i$ forward from its 1939-47 value to an identical value for 1947-54.

(B) Assume $\lambda_{i47} = \bar{\lambda}_{i39}$ for each industry. This means that the $\lambda_i$ for each industry for the 1947-54 period was simply the mean of the $\lambda_i$ for the prior period. In a sense this implies that the growth rates in the region are a fixed differential of the U. S. growth rate which does not differ by industry. This technique assumes only a regional effect on growth.

(C) Assume $\lambda_{i47} = 0$, for each industry. This means that the $\lambda_i$ for each industry for the 1947-54 period is 0. There is no difference between U. S. and New England growth rates in each industry. This also assumes that there is no influence played on current growth behavior by events of the past.

(D) Assume $\lambda_{i47} = \lambda_{i39} + \frac{1}{2} (\lambda_i - \bar{\lambda})_{39}$. This implies that in the subsequent period the $\lambda_i$ regress back to their 1939 mean value by one-half the difference between their mean and actual values. This is a variant of (B) above.

It should be pointed out that there are no *a priori* grounds on which to judge which *ad hoc* model should be used as a null hypothesis against our theoretical structure. We started with the first (A) in mind and then developed the others, when detailed examination of the data revealed a substantial degree of regression back to the mean among the $\lambda_i$ between one period and the next. There are no *a priori* reasons for expecting the $\lambda_i$ to regress to a common value in a subsequent period. We suspect that it occurred between the two periods under examination because of the economic events going on at the time. The first period encompasses the rapid return to full employment and the expansion of wartime activities. The consequence is that the first period represented a reutilization of capacity for wartime purposes that would have not been renewed in the long run. The second period represents mainly a return to prewar secular tendencies in a number of significant New England machinery sectors.

Tables 3, 4, and 5 summarize the errors of prediction using the regression techniques and then the alternative techniques. In Table 3 we see the values of $\lambda_i$ which the different techniques specify. These values of $\lambda_i$ were then combined with the U. S. growth rates to produce projections for each of the nineteen New England machinery industries for 1954. The projections are subtracted from actual New England employment in these industries in 1954, and an absolute per cent error is computed. The absolute per cent errors and their unweighted means are shown in Table 4.

A comparison of the regression technique with the *ad hoc* projections shows the former in a very favorable light. In Table 4 we see that the average absolute per cent error is 29 per cent for the regression technique. This compares with 60 per cent for simple extrapolation (A); with 21 per cent for extrapolation of $\bar{\lambda}_{39}$ (B); with 21 per cent when we assume no differences between regional growth rates (C); and 39 per cent when we assume the $\lambda_i$ regress halfway to their mean (D). A comparison of performance by individual industries for each technique is shown in Table 3.

It can be seen that regression is clearly superior to simple extrapolation; it fares better than the latter in fourteen out of nineteen cases. It fares slightly better (ten out of nineteen) than the prediction (D). It fares slightly worse than prediction (C) and quite badly relative to prediction (B). These are consistent with the average percentage errors shown previously.

It is worth noting that the superiority of Method B as a predictor is not as clear-cut as the previous comparison indicates. Both Method B and Method C yield the same average per cent error.

TABLE 3. VALUES OF $\lambda$ FROM VARIOUS PREDICTIVE TECHNIQUES, 19 MACHINERY INDUSTRIES IN NEW ENGLAND

| | 1 | 2 | 3 | 4 | 5 | 6 | 7 | 8 |
|---|---|---|---|---|---|---|---|---|
| Industry No. | $\hat{\lambda}$ from regression | Actual $\lambda$ 1939-47 | Actual $\lambda$ 1947-54 | $\hat{\lambda}_{47}$ regress halfway to mean* | $\hat{\lambda} - \lambda_{47}$ 1−3 | $\lambda_{39} - \lambda_{47}$ 2−3 | $\bar{\lambda}_{39} - \lambda_{47}$† | 4−3 |
| 3541 | − 38.57 | − 14.48 | + 1.43 | − 13.02 | − 40.00 | − 15.91 | − 12.99 | − 14.45 |
| 42 | − 35.25 | − 48.78 | − 21.01 | − 30.17 | − 14.24 | − 27.77 | + 9.45 | − 9.16 |
| 43 | − 18.81 | − 39.18 | − 32.29 | − 25.37 | + 13.48 | − 6.89 | + 20.73 | + 6.92 |
| 51 | − 80.50 | −110.95 | +229.51 | − 61.25 | −310.01 | −340.46 | −241.07 | −290.76 |
| 52 | − 23.56 | + 5.77 | − 13.64 | − 2.90 | − 9.92 | + 19.41 | + 2.08 | + 10.74 |
| 53 | + 84.44 | + 17.36 | + 42.16 | + 2.90 | + 42.28 | − 24.80 | − 53.72 | − 39.26 |
| 54 | − 16.41 | − 5.21 | + 1.38 | − 8.39 | − 17.79 | − 6.59 | − 12.94 | − 9.77 |
| 55 | − 4.11 | + 71.90 | − 10.57 | + 30.17 | + 6.46 | + 82.47 | − 0.99 | + 40.74 |
| 59 | − 25.50 | − 28.66 | − 8.56 | − 20.11 | − 16.94 | − 20.10 | − 3.00 | − 11.55 |
| 61 | + 31.76 | + 77.93 | − 28.08 | + 33.18 | + 59.84 | +106.01 | + 16.52 | + 61.26 |
| 62+63 | + 13.08 | − 75.08 | + 11.20 | − 43.32 | + 1.88 | − 86.28 | − 22.76 | − 54.52 |
| 66 | − 23.14 | − 85.33 | + 30.26 | − 48.44 | − 53.40 | −115.59 | − 41.82 | − 78.70 |
| 69 | − 0.16 | +105.11 | − 38.55 | + 46.77 | + 38.39 | +143.66 | + 26.99 | + 85.32 |
| 72 | − 1.95 | − 21.94 | − 1.16 | − 16.75 | − 0.79 | − 20.78 | − 10.40 | − 15.59 |
| 83 | − 21.57 | + 4.38 | + 12.28 | − 3.59 | − 33.85 | − 7.90 | − 23.84 | − 15.87 |
| 85 | − 23.76 | − 62.36 | + 22.01 | − 36.96 | − 45.77 | − 84.37 | − 33.57 | − 58.97 |
| 91 | − 17.58 | − 68.57 | + 9.46 | − 40.06 | − 27.04 | − 78.03 | − 21.02 | − 49.52 |
| 93 | − 5.92 | − 29.30 | − 15.15 | − 20.43 | + 9.23 | − 14.15 | + 3.59 | − 5.28 |
| 99 | + 13.93 | + 87.79 | + .96 | + 38.11 | + 12.97 | + 86.83 | − 12.52 | + 37.15 |

* This is the prediction $\lambda_{47} = \lambda_{39} + \frac{1}{2} (\lambda - \bar{\lambda})_{39}$

† $\bar{\lambda}_{39} = -11.5579$

TABLE 4. ABSOLUTE PER CENT ERRORS OF PREDICTION OF 1954 NEW ENGLAND EMPLOYMENT IN MACHINERY FROM VARIOUS TECHNIQUES

| | | (A) | (B) | (C) | (D) |
|---|---|---|---|---|---|
| Industry No. | From regression | $\lambda_{i47} = \lambda_{i39}$ | $\lambda_{i47} = \bar{\lambda}_{39}$ | $\lambda_{i47} = 0$ | $\lambda_{i47} = \lambda_{i39} - \frac{1}{2}(\lambda_i - \bar{\lambda}_{39})$ |
| 3541 | 34.16 | 13.59 | 11.09 | 1.22 | 12.34 |
| 42 | 16.34 | 31.87 | 10.82 | 24.11 | 10.52 |
| 43 | 12.81 | 6.54 | 19.69 | 30.68 | 6.58 |
| 51 | 97.05 | 100.00 | 75.39 | 71.77 | 91.00 |
| 52 | 18.15 | 35.51 | 3.80 | 24.95 | 19.65 |
| 53 | 36.44 | 21.40 | 46.31 | 36.39 | 33.89 |
| 54 | 19.95 | 7.38 | 14.51 | 1.55 | 10.96 |
| 55 | 8.32 | 106.11 | 1.30 | 13.59 | 52.43 |
| 59 | 10.59 | 14.47 | 6.55 | 20.75 | 3.97 |
| 61 | 75.38 | 133.57 | 20.81 | 35.37 | 77.17 |
| 62+63 | 1.37 | 64.59 | 17.08 | 8.35 | 40.77 |
| 66 | 43.74 | 94.69 | 34.25 | 24.79 | 64.47 |
| 69 | 55.04 | 259.05 | 32.93 | 55.31 | 145.99 |
| 72 | 1.16 | 30.50 | 15.26 | 1.70 | 22.89 |
| 83 | 36.26 | 8.47 | 25.52 | 13.15 | 16.98 |
| 85 | 37.61 | 69.33 | 27.59 | 18.09 | 48.45 |
| 91 | 26.10 | 75.34 | 17.45 | 5.89 | 45.95 |
| 93 | 11.51 | 17.65 | 4.48 | 18.90 | 6.59 |
| 99 | 7.76 | 51.93 | 7.48 | 0.57 | 22.22 |
| Average (%) | 28.93 | 60.10 | 20.65 | 21.43 | 38.57 |

TABLE 5. COMPARISON OF ERRORS OF PREDICTION FOR INDIVIDUAL PREDICTING TECHNIQUES FOR 19 INDUSTRIES

| Regression does | | (A) $\lambda_{47} = \lambda_{39}$ | (B) $\lambda_{47} = \bar{\lambda}_{39}$ | (C) $\lambda_{47} = 0$ | (D) $\lambda_{47} = \lambda_{39} + \frac{1}{2}(\lambda - \bar{\lambda})_{39}$ |
|---|---|---|---|---|---|
| better | | 14 | 4 | 9 | 10 |
| worse | | 5 | 15 | 10 | 9 |
| cases | | 19 | 19 | 19 | 19 |
| Average per cent error | 29 | 60 | 21 | 21 | 39 |

Further, Method C actually predicts better than B in eleven out of nineteen trials. The superiority of both methods over the regression equation technique will be discussed below.

### *The Consistency of our Data with the Original Hypotheses*

How consistent are the data with our predicting theory? The salient conclusions of our theory are:

1) The relative growth ($\lambda$) of a New England industry is determined by the relative return on investment ($\mu$).

2) Where an industry's growth is underestimated in Period 1, this growth will accelerate in Period 2, and vice versa.

3) Failure to predict $\lambda$ in Period 2 from the Period 1 regression would be due to changes in $\mu$ from one period to the next.

We have seen the first hypothesis confirmed by the good regression results between $\mu$ and $\lambda$ for the two time periods under examination. This is also shown by the relative success of the regression equation as a predictive device over simple extrapolation. The second hypothesis is also confirmed by the data; confirmed so well in fact that it almost looks as if the $\lambda_i$ change sign from one period to the next. Actually they change sign in eleven out of nineteen cases. What is important, however, is the confirmation that the regression relation indicated the change in the value of $\lambda_i$ by means of the error in the regression. It can be seen by comparing the residuals from the regression with the actual changes in $\lambda_i$ which occurred in the two periods. This is shown in the following tabulation.

| Industry No. | $\lambda_{47} - \lambda_{39}$ | $\hat{\lambda} - \lambda_{39}$ | | $\lambda_{47} - \lambda_{39}$ | $\hat{\lambda} - \lambda_{39}$ |
|---|---|---|---|---|---|
| 3541 | + 15.91 | −24.09 | 3562+63 | + 86.28 | + 88.16 |
| 42 | + 27.77 | +13.53 | 66 | +115.59 | + 62.19 |
| 43 | + 6.89 | +20.37 | 69 | −143.66 | −105.27 |
| 51 | +340.46 | +30.45 | 72 | + 20.78 | + 19.99 |
| 52 | − 19.41 | −29.33 | 83 | + 7.90 | − 25.95 |
| 53 | + 24.80 | +67.08 | 85 | + 84.37 | + 38.60 |
| 54 | + 6.59 | −11.20 | 91 | + 78.03 | + 50.99 |
| 55 | − 82.47 | −76.01 | 93 | + 14.15 | + 23.38 |
| 59 | + 20.10 | + 3.16 | 99 | − 86.83 | − 73.86 |
| 61 | 106.01 | −46.17 | | | |

The first column shows the change in $\lambda_i$ from one period to the next $(\lambda_{47} - \lambda_{39})$. The second column shows the residuals from the 1939-47 regression line. If our hypothesis is valid, we should observe the following: If the original residual from the regression, $(\hat{\lambda} - \lambda_{39})$, is positive, the change in $\lambda$ should be positive. If the initial residual is negative, the change in $\lambda$ should be negative. In other words, $\lambda_{47}$ moves away from $\lambda_{39}$ in the direction of the value predicted by the regression line. An examination of the table shows a very strong confirmation of this. In sixteen out of nineteen cases the signs agree, indicating that $\lambda_i$ does indeed move in the direction predicted by the regression.

The third hypothesis is also borne out by the data. The failure to predict $\lambda_{i47}$ from the 1939-47 regression line is closely related with the change in the $\mu_i$ from one period to the next. The following tabulation shows the error in predicting $\lambda_{47}$ from the regression $(\lambda_{47} - \hat{\lambda})$ and the change in $\mu_i$: $\Delta\mu_i = \mu_{47} - \mu_{39}$. The error from regression is here shown as the negative of that shown previously, to conform to the dating of the $\mu_i$.

| Industry No. | $\lambda_{47} - \hat{\lambda}$ | $\mu_{47} - \mu_{39}$ | | $\lambda_{47} - \hat{\lambda}$ | $\mu_{47} - \mu_{39}$ |
|---|---|---|---|---|---|
| 3541 | + 40.00 | +101.07 | 3562 + 63 | − 1.88 | −59.04 |
| 42 | + 14.24 | + 32.68 | 66 | +53.40 | +43.65 |
| 43 | − 13.48 | + 4.01 | 69 | −38.39 | −55.74 |
| 51 | +310.01 | +432.54 | 72 | + 0.79 | −69.60 |
| 52 | + 9.92 | + 16.60 | 83 | +33.85 | +32.94 |
| 53 | − 42.28 | − 6.18 | 85 | +45.77 | +75.51 |
| 54 | + 17.79 | − 6.20 | 91 | +27.04 | +99.29 |
| 55 | − 6.46 | − 33.69 | 93 | − 9.23 | −45.68 |
| 59 | + 16.94 | + 99.06 | 99 | −12.97 | −82.74 |
| 61 | − 59.84 | − 85.05 | | | |

The correlation between these two terms is +.9340. This indicates quite conclusively that the changes in $\lambda$ conform to our theory.

## *Conclusions*

We have presented a theory explaining industrial growth among regions which leads to predictions concerning relative growth rates.

We have shown that the relation between relative rate of return and relative employment growth may be used to predict relative employment growth in a region in a subsequent period. This relative employment growth may be used to project the trend of employment in a region, if projections of such trends on a national level are available and reliable.

The theory has been tested for predictive accuracy in terms of the available data and against *ad hoc* projections. Every implication of the theory is compatible with the data.

1) The relative growth ($\lambda$) of an industry is determined by the relative return on investment ($\mu$).

2) Where an industry's growth is underestimated in Period 1, this growth will be accelerated in Period 2, and vice versa.

3) Failure to predict $\lambda$ in Period 2 from the Period 1 regression would be due to changes in $\mu$ from one period to the next.

The tests show that only one *ad hoc* predictor was clearly superior to our theory. To be sure, there is no reason to expect that the same *ad hoc* method of prediction will work for another industry or for another period. On the other hand, the theory provides us with an unambiguous set of instructions on procedure and prediction, which has a rationale.

Finally, our theory is more *fruitful* than other predicting theories. We claim the ability to explain why the relative rate of growth *changes* from one period to the next. What other theory can make such a statement with as good results (cf. the above tables)?

## *III. APPLICATIONS OF THE MARGINAL EFFICIENCY CONCEPT TO SOCIAL POLICY*

In the preceding section the predictive aspects of the marginal efficiency concept were explained. In this section we shall apply the concept to the important policy issues in declining regions.

A declining region may be defined as an area in which the demand for labor is not shifting to the right as rapidly as the supply schedule of labor. The consequence is that those entrants into the labor market who do not wish to migrate are faced with employment opportunities yielding a lower rate of growth of wages than can be earned elsewhere. A number of public policies at the state and federal level have been proposed to improve this situation. Many observers believe that, left to themselves, the declining regions will continue their slow decay. However, there is disagreement over the direction policy should take.

It is clear that there is a loss of economic efficiency during the process of economic decline. The loss of efficiency consists in the difference between the maximum income the resources could produce under competitive equilibrium and the income they actually yield in their present employments. Many workers and businessmen would rather earn lower than maximal returns for their efforts in preference to migrating. The failure to migrate imposes additional losses on future members of the labor force.

Policy toward declining regions may take several forms. It may take the form of inducements to workers to leave the region and seek employment elsewhere. Alternatively, it may take the form of subsidizing the entry of capital into the region in an attempt to revive the sources of employment in the local economy.

From the standpoint of economic efficiency, the question of appropriate social policy may be phrased in the following way: Can the government expend a sum *equivalent* to the social long-run marginal costs of moving workers from Area A to Area B and thereby raise the real standard of living of the workers by more than it would be raised if they migrated? This expenditure would take the form of construction, subsidized loans, and tax rebates. It would be designed to increase capital in the depressed area, to generate jobs, and to raise the productivity of labor in existing jobs.

If the answer to this question is affirmative, then we would say that government intervention would increase efficiency by interfering with existing regional patterns of growth and decay. If the answer is negative, then efficiency is not served by intervention; and the rationale for intervention is solely the amelioration of distress caused by the existing patterns of growth and decay. This is no small consideration, but it is conceptually desirable to separate the efficiency argument from the social dislocation argument.

We propose to show how one may gauge whether in fact it is more efficient to:

1) invest social resources to transfer labor from depressed to growing areas; or

2) invest social resources in training workers and building plant and equipment in depressed areas. The answer depends on the following considerations:

a) The marginal efficiency of investment of potentially growing firms and industrial complexes in these declining regions.
b) The cost of moving workers out of the region.
c) The cost of social services workers consume in declining regions, compared to what they consume in expanding regions.
d) The increase in real standard of living a worker enjoys moving from declining to expanding regions.
e) The transfer wage of a worker.

While all of the above considerations enter the final answer, the crucial element is (a) above. Why should there be a greater return to investment in plant and equipment when it is carried out by a government authority than when it is carried out by a private business? Obviously, the two rates of return may equal each other and in turn equal a rate below the social cost of capital. If this is the case, intervention is not clearly justified on efficiency grounds. In order to justify intervention it is necessary to show either that social benefits are greater than private benefits or that social costs are less than private costs such that the *net* social rate of return exceeds zero. The net refers to the opportunity costs of investment. The argument usually given here is similar to that used by the sponsors of "infant industry" subsidies. Get our industry on its feet by subsidy and the increasing returns which its expansion will generate will make it self-sufficient in a number of years. Presumably, the increasing returns used in the argument are the economies of the division of labor which operate both internally to the firm and externally to the industry. While this argument has many attractions, it is difficult to quantify. All too often, investigators of location problems attribute everything they cannot explain to external economies. The argument we present below is quantifiable and therefore can be made with somewhat more precision and perhaps more justification.

It will be recalled from the preceding section that we computed a $\mu_i$ and $\lambda_i$ for each of nineteen machinery industries for the United States and New England. $\mu_i$ was the difference in rate of return, and $\lambda_i$ the difference in rate of growth of employment.

New England consists of a number of subregions undergoing divergent growth patterns. Southeastern New England (Rhode Island and southeastern Massachusetts) and northern New England (Maine, Vermont, New Hampshire) are growing less rapidly than the remainder, consisting of Connecticut and the rest of Massachusetts. If we were to compute an $m_i$ and $L_i$ for these subregions, we would certainly discover their $L_i$ lower than in the more rapidly growing parts of New England, and we might discover the same for their $m_i$.

In order to discover whether the computed $m_i$ of the declining sector is a true indicator of the return to social investment in the $i^{th}$ industry, we would have to explain the sectoral variation among the $m_i$ and $L_i$ for each New England industry. This has been attempted in other studies,[18] and the following influences emerge:

A) Sectoral differences in tax structure would appear to influence the $L_i$, for a given value of $m_i$. It will be recalled that the computed marginal rate of return was gross and included taxes and depreciation. Tax structure differences may nevertheless influence the location of a particular industry. For example, certain localities in New England impose the standard town property tax on the inventory and equipment of manufacturing corporations. Other localities do not. A difference of this sort, even when unaccompanied by differences in total tax revenue, would be of extreme importance in understanding differences in growth patterns.

B) The geographic concentration of an industry in one sector would appear to be an important influence on the $m_i$, the rate of

[18] See *Economic Factors in the Location of Industry,* Part One and Part Two, *op. cit.;* Stein and Schupack, *The New England Machinery Industry in 1970* (Federal Reserve Bank of Boston, 1960).

return. This is the way in which the external economies apparently play their role.[19] If a sector wished to stimulate the growth of an industry subject to such economies, it has two alternatives:

i) either lure away the larger, self-contained units which do not rely on such economics, or

ii) attempt to reproduce the industry in sufficient scale so that the economics reappear.

C) Differences in the average size of firm in an industry appear to be an influence on sectoral differences in $m_i$. In one sector, the firms may be large enough to enjoy certain internal economies of mass production, or of entry into capital markets. If another sector tries to stimulate growth in this industry, it will only be successful if it can induce growth on the part of the larger, more viable units.

D) Perhaps as important as the others, certainly in the short run in New England, is the vigor of entrepreneurship: This affects both the $m_i$ and the $L_i$. It affects the $m_i$ insofar as management refrains from choosing the most profitable opportunities for investment and growth. It affects the $L_i$ insofar as management fails to push investment forward as far as the rate of return would support. Most of the declining urban areas of New England based their original prosperity on textiles. The entrepreneurial driving force behind this industry is no longer present.[20] While it is true that the diverse immigrant groups which manned the textile mills have become Americanized, they have not yet become sufficiently imbued with the entrepreneurial drive. The presence of single-family enterprises, the lack of higher education, and the lack of contact with technological developments in manufacturing are strong barriers to the formation of large manufacturing enterprises, even by groups with substantial amounts of accumulated capital.

It is this last factor which is nevertheless the greatest hope of revival of depressed areas in New England. For it is this factor

[19] George J. Stigler, "The Division of Labor is Limited by the Extent of the Market," *The Journal of Political Economy*, 59 (June 1951), 185.

[20] This is stressed in G. H. Borts, *Economic Factors in the Location of Industry*.

which yields the possibility that high rates of return in a growing sector may be emulated in a declining sector. For entrepreneurship is a mobile factor. If local investors are irrational in their judgments of the profitability of investment in the southeastern New England area, prospects should be especially favorable for outside entrepreneurs and investors. In the long run they can be expected to come in to take advantage of these opportunties—if they exist. Only in the short run does irrationality prevail. But a lot of social distress can occur in the short run.

To conclude the argument: the desirability of social investment in declining regions hinges on the return to social capital as opposed to private capital. The return to social capital may be higher than the actually observed return to private capital for two reasons: One reason is the existence of external economies. The second reason is the failure of local entrepreneurship to perform as efficiently as market opportunities allow. The consequence is that the observed marginal efficiency of investment in a subregion is lower than the same industry is earning in neighboring subregions, or could earn in this region if the existing firms were more efficient. This problem would not exist if potential entrepreneurs and investors from other areas were unaffected in the short run by the pessimism of indigenous firms. In terms of our analysis, the $\lambda_i$ may not be as large as expected on the basis of the $\mu_i$ in these areas because of the noneconomic factors discussed earlier. Here again there may be an argument for social action to carry investment to the level which is indicated by the objective rather than the fancied rate of return.

E) Having formalized an answer to the desirability of social intervention, we must next examine the possibility that social intervention can be successful.

To what extent can a declining area be revived through the transplant of growing sectors of neighboring areas? Is revival possible, or would socially desirable investment become a permanent subsidy? It is one thing to argue that the return of social investment is high enough to justify it. It is another to argue that the social investment will stimulate a private economy to thrive without future subsidy. To take New England as an example, there are certain industries in the area that are growing more rapidly and have higher rates of return on investment than in the United States. If these industries could be stimulated to grow in the decaying sections of southeastern New England, such areas might be revived. For these are manufacturing industries, requiring no special soils or climate, and we know by inspection of the regions that they are sufficiently similar in terms of topography and labor force that either could support the industries in question. In other words, we seek to revive a subregion by emulating the growing characteristics of the whole area. Before we can do this, however, it is necessary to explain why this group of industries thrives in the whole region and does so poorly in the subregion. If we can quantify the differences in $m_i$ and $L_i$ between the whole region and the subregion in terms of a set of policy variables, then we can argue that transplanting can occur. For example, if changes in the tax structure or in certain types of social legislation would allow these industries to grow as well in a subregion, then the stimulation could occur. If the differences are more fundamental and rest on the internal economies of plant size and the external economies of geographic concentration, trouble in transplanting arises. It is not completely impossible to emulate the growth of the growing region, but it must be done selectively in terms of the characteristics which will allow the new industry to flourish. Finally, if the differences can be explained only by sociological variables, such as the decayed accumulation instincts of old families, then the transplants may be highly successful. The whole point of stimulating industrial growth will then be the creation of a new entrepreneurial group which can run and expand the subsidized industrial complex. In our studies of southern New England, this last phenomenon takes on major importance. It indicates that social investment to stimulate industrial revival must be large in magnitude and continued over substantial periods of time. This does not imply that such investment would have a low return, but it does indicate that investment in industrial revival is a long-run problem.

## IV. THE APPLICATION OF THE MARGINAL EFFICIENCY CONCEPT TO REGIONAL ACCOUNTS

Our use of the marginal efficiency concept has been limited by the available data. In some cases these limitations have severely restricted the full utilization of the predictive model.

1. Our choice of the years 1939, 1947, and 1954 to illustrate the concept is made necessary by available Census data. It would be preferable to be able to observe our variables in years which more adequately reflect the long-term growth of the different regions. At the beginning of the paper it was argued that years that are the peaks of the business cycle provide good terminal dates for the measurement of growth trends. Changes between these dates reflect secular forces rather than the intervening fluctuations of the business cycle. The three years 1939, 1947, and 1954 are poor choices from this standpoint, both 1939 and 1954 being trough dates of different cycles. That growth occurs at all between the last two dates is due to the weakness of the 1953-54 recession relative to the expansion of the preceding five-year period.

One obvious improvement in data would be the availability of the information for a greater number of years. This will permit the use of some discretion in the choice of years to use as terminal dates in the measurement of growth.

2. One of the greatest limitations imposed by the data is the lack of information on net investment in fixed and in working capital. The Census sources we employed provided only a figure which is the equivalent of gross investment in plant and equipment. There is no information on the retirement of property or on the depreciation of equipment to allow a net investment to be computed. While we have been able to make use of the marginal efficiency of gross investment as a predictor, it would be desirable to have the net investment data for all forms of assets. This would permit us to perfect our device without the use of proxy measures of net investment. It would also permit a more accurate identifi-

cation of subregions and sections of industries which were growing.

3. The lack of data on depreciation (i.e., required replacement) expense also hinders the computation of entrepreneurial income. The measure we derive from Census sources includes depreciation. To identify the net change in income, it would be desirable to exclude it.

4. One of the crucial pieces of information in the analysis of investment has to do with taxes. The information required is the volume of tax payments by type of tax and type of collector. This information is rarely included with Manufacturing Census data, which is presented on an establishment basis. Consequently, an attempt to provide such data would run into serious questions of organization of Census information. Nevertheless, in a catalog of missing pieces of knowledge, taxes take a prominent place.

5. Finally, our entrepreneurial income figure as constructed from Census data includes inventory decumulation. This arises from the fact that the Census reports the value of shipments instead of the value of output of each sector of an industry. The two concepts differ by the amount of inventory accumulation. To the extent that this accumulation takes on an uneven regional pattern, it introduces an error into our measurement of entrepreneurial income and its changes. It would be desirable to know the value of output of establishments in geographic sectors of the same industry to eliminate this source of error.

In addition to these limitations we might point out certain types of data which would not be of particular use in the evaluation of marginal efficiencies. Many investigators have proposed the revival of a census of wealth by regions. Due to the theoretical difficulty in valuing income-earning assets, we feel that such information does not materially contribute to the analysis of regional investment. Investment represents the amount of national product that is added to the stock of capital in any time period. Due to changes in valuation, this may or may not coincide with an increase in the value of capital. Because capital is essentially measured in value terms, it really doesn't contribute to the study of growth or growth differences among regions. What we need to pursue the analysis of regional growth is better data on investment and entrepreneurial income, for more dates and more industries.

## COMMENT

*Daniel Creamer*

*National Industrial Conference Board*

My prepared comments made at the Conference on Regional Accounts were largely directed at the data deficiencies that had not been recognized by Professors Borts and Stein and the probable effect on their projections. Since the paper in the present and revised version incorporates a number of these points, with gracious acknowledgment, nothing would be gained by repeating the same remarks for publication. Rather, I would like to comment on two points on which we still seem to differ. One relates to a technical consideration in methodology, and the other concerns the merits claimed for this approach by the authors.

Depreciation is high on the list of data they want developed. They rightly point out this would enable them to work with net capital expenditures instead of gross expenditures and would also bring them closer to the figure of net proprietary income. They also mention data they *do not* want, specifically census of wealth by regions. As they explain, "Investment represents the amount of national product that is added to the stock of capital in any time period. Due to changes in valuation, this may or may not coincide with an increase in the value of capital. . . . Because capital is essentially measured in value terms, it really doesn't contribute to the study of growth or growth differences among regions." The authors overlook the point that depreciation is also measured in the same value terms. If one is able to cope with depreciation figures, one is also prepared to deal with figures on capital stocks.

Moreover, there is a technique for adjusting for the price changes incorporated in the value of capital, and it has yielded reasonable results for many types of analysis. The existence of capital estimates could serve as a check on the estimates of capital expenditures which, by their very nature, are bound to be subject to relatively large errors of estimate.

Professors Borts and Stein conclude that their "theory is more *fruitful* than other predicting theories. We claim the ability to explain why the relative rate of growth *changes* from one period to the next. What other theory can make such a statement with as good results?" [Emphases in the original.] The authors themselves have shown the connection between differentials in rates of profit, volume of investment, and employment. Therefore changes in employment, which is referred to as an *ad hoc* predictor, can be traced to changes in differential rates of profit. Thus, both predictors have the same rationale, and one variant of the *ad hoc* predictor actually proved to be a superior predictor in the one test performed by Messrs. Borts and Stein.

However, to be truly fruitful, I should like to suggest that it is necessary to identify the elements that result in a higher rate of profitability in a given industry and the impact of regional differences on the strength of these elements. This insight is not provided by the predictor recommended by Professors Borts and Stein, nor by the best of their *ad hoc* predictors. Stated another way, in the best of circumstances extrapolation is an inadequate substitute for derivation. In brief, there is no short-cutting the more difficult task of industry studies in the framework of locational economics.

## *Richard F. Muth*

*University of Chicago*

Borts and Stein have given us a truly fine paper, one which is very suggestive of why industries grow at different rates in the various subnational sectors of our economy. There are many things that I like about their paper. It deals with a substantive problem, uses traditional economic analysis in an attempt to explain an old, but as yet unresolved, question, and presents a meaningful hypothesis—one which could conceivably be refuted by empirical evidence. However, I shall argue that the Borts-Stein theory is incomplete and that the evidence for the use of their technique in forecasting regional growth rates is not as favorable as they seem to believe.

The hypothesis advanced in this paper is that differential changes in employment are positively related to differences in the marginal efficiency of investment. The authors consider the case in which an industry is composed of several sectors, all of which sell in a common market. If the demand for the industry's output increases, the marginal efficiency of investment and of employment both increase most rapidly in the sector where the elasticity of labor supply is greatest. However, not all groups of firms labeled an industry by the Census Bureau sell in a common market. When we consider firms such as those in the bakery products group, to take an obvious example, the possibility arises of differential shifts in the demand schedules facing firms in different regional sectors. Likewise, it is not inconceivable that differential shifts in the supply of various productive factors would lead to differential changes in employment among regions. In particular, if the supply of capital were to increase more rapidly in some sectors than in others, we might actually find a negative association between employment changes and capital earnings. Thus, I would not be surprised to find somewhat different relationships between employment changes and differences in the marginal return to capital if data for other industries or time periods were examined.

Turning now to the empirical evidence presented by Borts and Stein, two sets of considerations are relevant: the empirical confirmation of their hypothesis, and the evidence bearing on its usefulness for forecasting. Clearly the evidence presented fails to contradict the hypothesis that employment growth in a sector varies directly with the marginal efficiency of capital. However, the evidence bearing on the usefulness of the Borts-Stein technique for forecasting is not nearly so favorable, in my opinion. It is certainly gratifying that the regression equation for 1939-47 yields better predictions of relative employment growth in the subsequent period than does naive Hypothesis A, which asserts that industries growing relatively more rapidly in New England in the past will continue to do so. However, naive Hypothesis C, which states that any given industry will grow as rapidly in New England as in the nation as a whole, is simply the basis for the weighting of national growth trends, a procedure which the authors show to be so unsatisfactory earlier in their paper. While Borts and Stein did not consider this alternative until after examining the data, it is certainly a natural one for comparison with their procedure, in view of its wide use for forecasting. I therefore find it disappointing that the technique developed here does not do substantially better instead of somewhat worse than naive-C.

# 4

# IMPACT OF LOCAL GOVERNMENT POLICIES ON LOCAL CAPITAL FORMATION AND OUTPUT

*Edward B. Berman*[1]
*Economic Factors Office, The Mitre Corporation*

THIS PAPER represents an elaboration of portions of the general model presented by Werner Hirsch at this Conference.[2] We shall attempt to set forth a conceptual framework for tracing the impact of local government policy decisions on the local economy. The independent variables in the analysis are the local government policy decisions themselves—tax structure and levels, zoning regulations, and so forth, all translated into their first order impacts on local industry. These operate on the local economy through a structure which is represented by means of a regional interindustry model and ancillary capital formation and export models. The output of the analysis is a set of impacts on the local economy measured in terms of production levels, capital formation, and the balance of trade with the extra-local world, all in commodity or industry detail.

[1] The author wishes to express his appreciation to Frederick T. Moore and Leon Moses for their constructive suggestions made as discussants at the St. Louis Conference on Regional Accounts.

[2] See Werner Z. Hirsch, "A General Structure for Regional Economic Analysis" in this volume, pp. 1-32.

The analysis is designed from the point of view of the local government decision-maker who wishes to compare the economic impacts of alternative sets of policies in order to choose that policy which will provide the greatest stimulation to the growth of the local region. For the purposes of our analysis, we assume that the set of local government policy decisions is translatable into a level of subsidy (positive or negative) for each industry relative to the initial costs of production of that industry. Thus, whether the policy decision operates through an actual subsidy, such as an offer to refund property taxes for the first three years of operation of a new plant; or whether the policy decision operates by lowering the expected costs of production, such as a favorable rezoning of land or mosquito control; in either case the effect is translated into a subsidy relative to the initial expected costs of production.

For the purposes of this analysis, we abstract from the difficult problem of translating local government decisions into their first-order impacts, the equivalent subsidies for each industry. We also ignore the difficulties in evaluating the welfare implications of the complex set of impacts which the model will generate. Our purpose is to design a structure within which it will be possible to trace the interactions of private decisions on each other, directly and through their economic implications. This does not imply that the preparation of inputs and the evaluation of outputs are not difficult and important tasks, but only that our interest at the moment is elsewhere.

We make no assumptions at the moment about the level of optimization at which the local government operates. It may, for example, wish to find the best policy possible for the local area, the state, the country, or the world; or perhaps the best policy possible for the local area subject to the constraint that the rest of the state, country, or world should not lose more than the local area gains. These are essentially moral problems in evaluating the outputs of the model which follows, and for now we avoid them.

## I. THE STRUCTURE OF INTERACTION

The model which represents the interaction between local government policy and the local economy is for the most part a rigid structure of interindustry coefficients. On the other hand, there are two points at which relatively complex decision models have been designed: (1) the capital formation process and (2) the competition for export markets. These decision models determine the corresponding final demand sectors for the regional interindustry model.

The model contains two or more time periods:[3] (1) one or more periods for building capital plant and equipment, each perhaps a year or two in length (the length should represent a compromise of the capital lead times of the industries in the local area); and (2) a level-off period. In the capital-building periods, requirements for the construction of new plant and equipment are generated by the model; in the level-off period, capital requirements are stipulated. For the purposes of the presentation to follow, we limit the model to one capital-building period and one level-off period. This limitation permits the use of the simple geometric analysis of Figure 1. In the three or more period case, the subsidy requirement, although equally deterministic, would become a function of two or more variables, namely, production required in the level-off period and in each capital-building period beyond the first.

### *The Capital Formation Decision Model*

The capital formation decision for a local industry is described in a model which represents the decision for a single local industry as a discrete function of (1) the requirements for local production and (2) the totality of local government policy translated into a subsidy level (not necessarily positive).

Figure 1 illustrates a typical capital formation decision model. The heavy saw-toothed line represents the negative of the present value of plants of the industry built in the local area as a function of the requirement for production by the local industry. Thus as

[3] This paragraph has been added at the suggestion of Frederick T. Moore.

required production begins to exceed existing capacity, the present value of an additional plant rises from a large negative value toward or beyond zero. In Figure 1, since we are measuring the negative of present value, the negative present value falls from *E* to *F*. Point *F* represents the optimal production level in the potential new plant; if the new plant is not worth building at this level of required production it will not be built at any other level. Thus, if the present value of the plant remains negative at Point *F*, as it does in Figure 1, a subsidy is necessary for the plant to be built. This subsidy level is the minimum subsidy level that will have any effect on this industry; if that amount of subsidy is not forthcoming the industry will import all excess requirements over capacity.

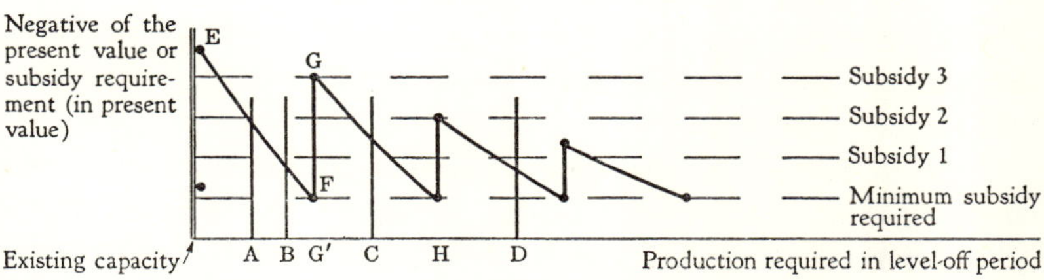


*Figure 1.* The capital formation model for a hypothetical local industry.

The kink in the negative present value curve at Point *F* represents a shift from consideration of the first new plant to consideration of the second new plant. Point *G* is lower than Point *E*, the corresponding point for the first plant, since we assume that production is spread evenly over all existing plants, including new ones. The excess capacity which would exist at Point *G* is therefore spread over more plants than would be the excess capacity at Point *E*, and the present value of the incremental plant is higher as a result. On the other hand, if the industry represents a single multiple-plant firm, the loss in production at existing plants will be charged against the construction of the new plant, and hence Point *G* will be equal to Point *E*.

We are then able to use the function of Figure 1, together with (1) the level of subsidy forthcoming as the output of a complex of local government policy decisions, and (2) the quantity of local production required in the level-off period to determine if any plants are to be built in this local industry, and if so, how many. Thus if local government policy is equivalent to Subsidy 1 in Figure 1, then a local production requirement of *A* (over existing capacity) would indicate that no plants will be built and the quantity *A* will be imported. At a local production requirement of *B*, one plant would be built and there will be no imports of the subject commodity. At a local production requirement of *C*, one plant will be built and the excess requirement (*C–G'*) will be imported. At a local production requirement of *D*, three plants will be built and there will be no imports of the subject commodity. If, on the other hand, local government policy constituted the equivalent of Subsidy 2 in Figure 1, one plant would be built at a production requirement of *A* and two would be built at a production requirement of *C*.

Through the use of time-phased capital coefficients (i.e., coefficients for the expansion of capital capacity), the outputs of the capital formation models can be translated into the construction and producers' durables sectors of final demand for the local industries under study in the capital-building period. To these final demand sectors should be added the simple product of the vector of time-phased capital coefficients times the requirements for new capacity in the level-off period (required local production minus initial capacity) for national industries located in the region and for any local industries which do not require the discrete analysis by reason of a plant size that is small relative to the size of the local industry.

### *The Export Decision Model*

It is possible to combine the capital formation model of Figure 1 with an analysis of the interrelationship between local governmental policy and exports to be forthcoming from the local industry. Exports can be introduced into the capital formation analysis in two ways. If the first export market that could be obtained, in order of the subsidy necessary to penetrate the market, were large enough to make a national industry out of the local industry, it would be sufficient to insert a maximum subsidy level in the capital formation model. Thus, if Subsidy 3 in Figure 1 represented the level of subsidy at which the hypothetical local industry penetrated a major market and became a national industry, we should be able to abandon the discrete capital formation analysis at a subsidy level equal to or greater than that. We should then move to an analysis of the level of exports, as shown in Figure 2.

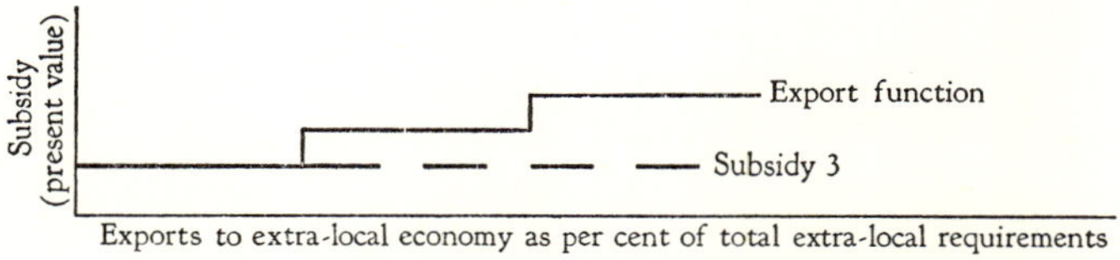


*Figure 2.* Export analysis for hypothetical national industry located in subject region.

In this first case there is no overlap between the capital formation analysis and the export analysis. At a subsidy level less than Subsidy 3, the industry is local; the discrete capital formation analysis is relevant but there will be no exports. At a subsidy level greater than Subsidy 3, the industry becomes national; the export analysis becomes relevant while the discrete capital formation analysis ceases to be needed. Given the level of subsidy-equivalence implied by local government policy decisions, the level of exports of the subject industry in terms of share of the extra-local market can be determined from the export function of Figure 2.

In the second case, in which the first export market penetrated is not sufficient to transform the industry into a national industry, it would be necessary to continue analyzing capital formation discretely as in the model of Figure 1. The level of exports implied from local government policy decisions can be determined from the export function and added into the discrete capital analysis.

In either of the two cases described above, it is necessary to assume that the extra-local world will remain politically passive.

If a local government policy of subsidizing the entry into new export markets were to be met by reactive shifts in policy by other local government units, a game-theory analysis would be required to estimate the increase in exports, if any, that would finally result from any given subsidy. We assume that such political reactions do not occur.

There is an implicit assumption in Figure 2 that the geographical distribution of extra-local requirements will remain constant as the total level of requirements is raised or lowered. This is a poor assumption, but an unavoidable one as long as we are not willing to abandon the regional analysis for an interregional analysis.

### *The Regional Interindustry Model*

The third element in the conceptual structure of interaction is a modified regional interindustry model.[4]

The typical regional interindustry model requires extensive exogenous column and row sectors, as depicted in Figure 3.

Sector *b* of the matrix of Figure 3 represents the impacts of the extra-local economy on the local economy. Here each coefficient

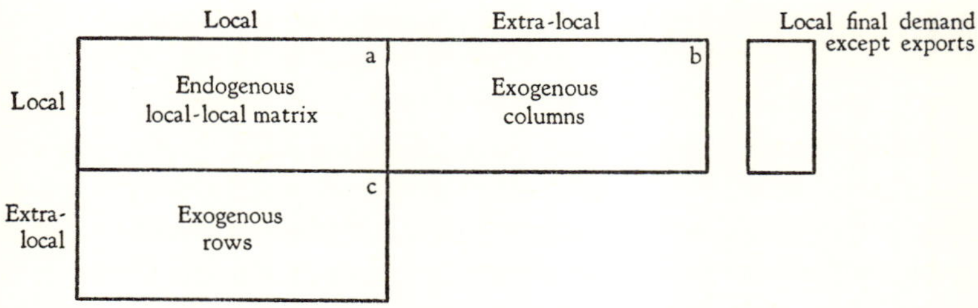


*Figure 3.* A typical regional interindustry model.

[4] For examples of regional interindustry models see F. T. Moore and J. W. Petersen, "Regional Analysis: An Interindustry Model of Utah," *The Review of Economics and Statistics,* 37 (Nov. 1955) 368-83; W. Hochwald *et al., Local Impact of Foreign Trade* (Washington: National Planning Association, 1960); and W. Z. Hirsch, *Projecting Economic Activity and Population of the St. Louis Metropolitan Area: The Input-Output Approach* (St. Louis: Metropolitan St. Louis Survey, 1958).

represents the requirement from the local industry (the row) per unit of output of the national industry (the column). Multiplying a set of extra-local production levels by this submatrix converts them into exports required from the local region and thus forms a part of final demand for the local-local matrix (Submatrix *a*). Submatrix *c* translates the local production levels determined by the model into local imports and thus determines the local impact on the extra-local economy. In our modified regional interindustry model, the export analysis as depicted in Figure 3 replaces the exogenous columns (Submatrix *b*) as the determinant of the local share of extra-local markets.

A special treatment is required for local industries in which the marginal production is to be imported. For each such industry, a special vector should be added to the local final demand representing the inputs required for production at full (present or planned) capacity. This can be obtained as the product of the column of input coefficients for the industry multiplied by the full capacity production level. The column of input coefficients should then be replaced in the matrix of Figure 3 by a column representing the importation of the subject commodity. This column will be null in the local-local submatrix (Submatrix *a* of Figure 3) but will have a one in the appropriate exogenous row representing the fact that all marginal requirements are to be imported.

## *II. THE INTERACTION SYSTEM IN OPERATION*

The interaction structure, consisting of the capital formation models, the export models, and the regional interindustry model, is supplemented in operation by a national interindustry model. The system is solved iteratively.

In order to operate the system, the following inputs are required:

1) a set of local government policies translated into a subsidy level for each industry in each period;

2) a final demand for the local region for each period, excluding exports for both periods and capital formation in the first period; and

3) a final demand for each period for the nation outside of the local region, excluding only exports to the local region (local imports).

The solution of the system begins with the export analysis which is the only part of the process not requiring an iterative solution. In the export analysis, the share of the extra-local market in each period is determined as a function of the local subsidy in the period (see Figure 2).

*The Iterative Cycle*

*Step 1.* The first step in the iterative cycle is the determination of extra-local production levels for each period, using the extra-local interindustry model and the extra-local final demand vectors. In the first pass, some estimate of local imports should be added to the extra-local final demand vectors (as exports). This could be base year local imports or any other estimate of local imports. The output of this step is a set of extra-local production requirements for each period.

*Step 2.* The second step in the iterative cycle is the determination of the vectors of local exports given the share-of-the-market analysis and given the vectors of extra-local production levels of Step 1.

*Step 3.* The third step is the determination of new capital formation given the subsidy levels, the local production requirements in the level-off period from Step 4 of the previous iteration, and the new export vector for the level-off period derived in Step 2. This step is omitted in the first pass. In using the new export vector, the net increase (or decrease) in exports over the previous iteration is added to (or subtracted from) the local production requirements since these production requirements already reflect the export vector of the previous iteration.

The total requirement for local production for an industry in which the marginal requirement is imported should be estimated as the sum of imports plus full local capacity.

*Step 4.* Given the export vector of Step 2, and the capital formation vector of Step 3, plus local final demand, local production levels in each period are determined through the use of the local-local submatrix of Figure 3 (or its inverse).[5]

*Step 5.* The exogenous rows of Figure 3 are multiplied by the local production levels to determine local imports. These become inputs into Step 1 and the process is repeated until convergence is obtained.

It is expected that this system would exhibit the strong tendencies towards convergence that are characteristic of interindustry models.

## COMMENT

*Frederick T. Moore*

*The RAND Corporation*

Mr. Berman has sketched out a model whose purpose is to show the impact of local government policies on local capital formation and output. I emphasize the phrase "sketched out" since the paper does not give a complete description of the model with full details of all of the elements and the way these elements are expected to interact.

My first comment concerns the capital formation model. It is a simple two-time-period model and that may hide a particular class of problems. There is nothing, it seems to me, to prevent the model from exploding or else generating excessively high import levels. For example, the final demand may generate a requirement for output greater than capacity in one industry (for all other industries capacity being sufficient to meet demand). But to build the additional capacity in industry $i$ may be just enough to increase required output above capacity in industry $j$, and to try to patch

[5] Since the matrix must be adjusted in each iteration to cover industries in which the marginal requirements are imported, in that iteration it would not seem efficient to use an inverse.

that up may put industry $k$ in trouble. The only alternative to this "explosion" of the model is to increase imports (or lower final demand). The two-time-period model hides the process of adjustment (e.g., building capacity in period $t–k$ so as to be prepared in period $t$).

Second, the capital formation model is an ingenious attempt to bring in the effects of price changes on the present value of plant, and hence on the profitability of building new capacity, in a local industry. But the model has a number of troublesome aspects. I doubt that the present value of plant can be directly related to production required (as Berman does) even with a saw-toothed type of function such as Berman uses in Figure 1 in his paper. The present value of a plant depends upon product prices in the future and not directly on production. Thus only if there is an explicit relationship between production and product prices can there then be a relationship shown between present value and production required. Even so, why should the segments of the saw-toothed chart be linear? Berman also talks about a single point as representing "the optimal production level in the potential new plant." What scale of plant is implied? Is it the most efficient plant that could be built in the long run, and is the point he refers to the optimum *optimorum* on the long-run cost curve? Also, since this is a sophisticated analysis of plant costs, it seems odd to then use this model in conjunction with a typical input-output model which assumes that all activities are infinitely divisible (i.e., that costs are constant for any scale of operation). Finally, in the capital formation model, if a subsidy is to be paid is it to be paid simply to the new plants or also to the old plants? If paid only to the new plants, then a differential cost structure in the industry will occur. I don't know how that effect can be accommodated in this particular model.

Third, so far as the export decision model is concerned I have only one comment to offer. Berman talks of a subsidy which would "make a national industry out of the local industry . . ." By this I assume he means that there is some level of subsidy which in one step converts a purely local industry into a national industry. But this is unrealistic. As an industry attempts to reach out into more distant markets, different costs and prices (and hence, perhaps, a varying subsidy) will be met. The problem of transforming local industry into national industry is not met efficiently by assuming that there is some single level of subsidy which would permit this transformation. For certain kinds of industries this type of transformation is technically infeasible.

## *Leon N. Moses*

*Northwestern University*

There is considerable interest among social scientists in the factors that cause certain areas of a nation to experience low or negative rates of growth. There is a corresponding interest on the part of government officials in doing something about these depressed areas. Mr. Berman has succeeded in pointing the way to what may prove a fruitful approach to the analysis of regional growth and policies for influencing it. However, aside from its guidepost function Mr. Berman's paper does not serve significantly the needs of theorists or policy-makers. He rejects or solves by assumption all of the welfare, most of the efficiency, and a good many of the general equilibrium issues involved in his model. On the other hand, the inputs into the model, its objectives, and formal workings are incompletely formulated and likely to provide policy-makers with more frustration than help. Nevertheless, the author has pointed to an original and interesting approach. The comments that appear below may prove useful should he decide to push the work further. Unfortunately, space restricts me to a few such comments.

Mr. Berman exhibits an awareness that his problem involves optimization. Thus, he states that the model will allow local decision-makers to choose *that* policy which will provide the *greatest* stimulation to the local economy. Yet, given the range of possible levels and kinds of subsidies, the numerous investment choices for existing or new industries, the range of possible types of social overhead investment, etc., it is a formidable task to choose among the "endless" variety of programs. Two types of changes seem to be in order. First, the scope of the problem should be narrowed: possibly the author should restrict himself to subsidies that reduce the cost of direct inputs. Second, this narrower problem should then be cast in formal programming terms. Such formulation may clarify aspects of the model that are now quite vague; i.e., what is being maximized, what the constraints are in terms of total subsidy to be allocated, quantities of immobile factors available in the local area, quantities of mobile factors that may be attracted to the area, demand and costs in other areas, etc.

The paper does not make clear whether the author is dealing with an infant-industry type case, or one in which a "permanent" subsidy is required. This issue cannot be dismissed on the ground that it involves welfare consideration, though, of course, it does. Local decision-makers will surely want to know when or the conditions under which the stimulated industries will no longer require a subsidy. They might even want to know what the redistributive effects of a subsidy would be. Mr. Berman's model does not formally link the level of subsidy required to generate investment with the rate at which the rest of the economy is producing and carrying out capacity-increasing and cost-reducing investments. His vague reference to a share of market analysis and his unexplained subsidy-export function are not enough.

Finally, and unfairly, I find it difficult to accept Mr. Berman's lack of interest in the distorting effects of numerous, local subsidy programs. Some alternative approach such as a federal-state program for retraining labor, and assisting it to migrate, should at least have been mentioned.

# 5

# REGIONAL BREAKDOWNS OF NATIONAL ECONOMIC ACCOUNTS

*Richard Ruggles and Nancy D. Ruggles*

*Yale University*

IN THE PAST ten years there has been a considerable increase in economic analysis focused at the level of the region. Many economists working at the regional level have generated ingenious and informative kinds of data to aid them in their analysis. Others have felt the need for a more general and comprehensive framework and have borrowed extensively from the national economic accounts. It is proper, therefore, to ask whether the national economic accounts can provide the kind of information needed for regional economic analysis and whether it would be fruitful to provide regional detail in our national data for these purposes.

This paper will not attempt to explore the informational requirements for economic analysis at the regional level. Instead, it will focus on whether regional breakdowns of the national economic accounts are desirable and, if desirable, whether they are feasible either independently or as a derivative in the process of obtaining the national estimates. Regional data might be produced centrally by those who produce the national statistics, or it might be produced on a decentralized basis in regions where the information is desired. It is, of course, not possible to make an unequivocal choice between these methods. On the one hand, no one would suggest that each region provide its own population data, or other information which is already provided on a centralized basis. At the other extreme, it is also obvious that a national office which was required to service all of the information needs for every region and municipality would be faced with an almost impossible task, not only because of the resources required, but also because a national office might be unable to obtain the local co-operation necessary to elicit certain kinds of information, in circumstances where a locally manned effort might be successful. Even though there is no categorical answer to this major question of centralization vs. decentralization, however, it should be quite possible to arrive at some judgments concerning the advisability of centralizing and decentralizing the provision of certain kinds of data.

In considering whether a detailed breakdown of national economic accounts would be useful for regional economic analysis, it will be important, first, to bear in mind what kinds of economic analysis are required at the regional level. The following section of this paper will outline briefly what the needs are. It should be remembered, however, that this over-all view is not that of a regional expert, but rather that of an outsider trying to get some idea of what is going on in this area. Second, it will also be necessary to consider what constitutes a region for the purposes of regional analysis. In many cases the key to the question of whether regional accounts can be derived as breakdowns of the national accounts will depend upon the definition of the region. Third, it will be useful to examine the nature of the national economic accounts, to lay a foundation for considering their applicability to regions. Given these steps, it should be possible to examine what kinds of breakdowns of the various kinds of national economic accounts are desirable and possible. Although this rather detailed examination is not intended by itself to serve as a blueprint for future statistical work, it may lead to general conclusions which will be pertinent in orienting the development of regional economic accounts.

## *I. THE USES OF REGIONAL DATA*

Regional economic analysis has been for the most part policy oriented. Those supplying funds for such analysis hope to obtain from the research guidelines for policy and information which will enable them to arrive at sensible decisions.

One of the most frequent aims of regional investigations is the development of policy to meet public administrative needs and to assess various possibilities for action. Such analysis is likely to be of a fiscal and financial nature. Thus, the estimation of potential tax revenues and the evaluation of expenditure needs are central to the undertaking. In its most primitive form this kind of analysis appears in the budgeting of the state and local governments, where it is undertaken without regard to private economic activity in the region. Where some additional resources are available for research in this area, the analysis is likely to include some exploration of the relation of the state or local government to private economic activity. Such a review is very important if one wishes to evaluate, for example, the revenue that might be obtained by a sales tax, an income tax, or a corporation tax, where none of these exists at a particular political level. In order to program a budget effectively, considerable knowledge is therefore required about the fiscal, financial, and income structures of both the public and private sectors and how these are related to each other.

In every intensive regional economic analysis, however, the work soon develops beyond this, into a comparative economic analysis of the particular region in question with other similar regions. For example, it is of interest to know whether a given locality is levying a heavier or lighter property tax than other communities surrounding it. Such a question can only be answered if data are available for a number of regions on the same basis. Recent studies of metropolitan areas have emphasized the fact that comparative economic analysis is one of the major tools of the regional economist. Differences in productivity, levels of income, labor force participation, and amounts of public services provided can all be evaluated on a comparative basis, and such analysis may be very useful in revealing areas suitable for policy action.

The programming of policy also requires forecasts and projections of the future at the regional level. This may require an analysis of the effect of national trends on the region in question, and of the impact of outside events. In addition, it will also be necessary to show how factors operating in a specific region may cause this region to develop differently from other regions. Thus, for example, the migration of people out of one region and into another region must be taken into account in projections for either region. It may even be that projections for the nation as a whole eventually may best be built up by studying regional projections and their interactions.

Finally, regional analysis is very much concerned with investment and the future growth pattern which such investment implies. Long-range planning requires a reshaping of the region's economy to bring it into line with changing national needs and with the resources available in the region. Again, the interplay between the regional level and the national level must be brought into the analysis; in this case attention must be focused upon changing industrial structures and investment capabilities, both regionally and nationally.

## *II. DEFINITION OF REGION*

Thus far in the discussion "region" has been referred to in a rather vague manner. There are two characteristics that can be used to define meaningful geographical areas. First, one may be interested in a specific geographic area because it conforms to a political subdivision. Such a region must formulate a line of political action with respect to a number of issues under its jurisdiction. This will give rise to the need for regional analysis. Second, a region may be defined in economic terms, containing several political units which are bound together by their common economic interest. Thus, in the case of a metropolitan area, a number of local communities may be involved in the region, but because of the economic unity of the area, it is useful to treat it as a separate region and as an entity in its own right, despite the lack of political unity. The economic definition of a region can also be applied to regions larger than metropolitan areas. A given section of the country may find itself subject to similar economic forces, and a general analysis of these forces for the section as a whole may be desirable. Thus, for example, the Federal Reserve Bank of Boston has carried out economic analysis of the New England area in an effort to develop policies which would strengthen its growth and make up for the decline of the textile, leather, and non-ferrous metals industries.

There is no ready answer to what the smallest regional unit should be. Political and economic criteria may, in many cases, lead to entirely different regional breakdowns. In any case, it seems reasonable to aim for breakdowns by states and local political subdivisions, and in addition by important metropolitan areas. Data for the larger economic regions, such as New England, the Southwest, etc., could be built up from data for states, but this may not be true of metropolitan areas, given the complex characteristics. These regions therefore are likely to require separate identification.

Besides the geographical coverage of a region, it is also necessary to decide what kinds of data are to be covered within the region. Specifically, should the data about a region pertain to the economic activity taking place within the area, or should it cover the economic activity of the residents living within the area? In the field of national economic accounts this problem has been raised many times, and many of the finer points of national income accounting are concerned with it. A favorite topic of conversation is whether the pay of, say, a Norwegian sailor on a Dutch freighter registered in Panama en route between New York and South America, should be accounted in the national income of Panama, Holland, Norway, or even the United States. Most generally, the details on the national level have been worked out and somewhat arbitrary rules have been set up for classifying transactions between various countries.

The problems are multiplied, however, when we consider accounts for regions. Commuters may be treated the same as border workers between countries, and nonresident owners of property may be treated the same as foreigners holding private property in another country, but other problems are somewhat more difficult. Are the purchases made by people in New York City, other than those of residents of the metropolitan area, to be treated as tourism, as they are in the national accounts? Can national corporations whose legal place of incorporation is outside the region be treated as foreign companies, and the operating surplus of the plants within the region be treated as a branch profit? Many insurance companies have their home office in one part of the country, but to say that all their collections and disbursements are financial transactions occurring within that particular region may seriously distort the economic realities. Most of these questions as to the transactions to be covered will have to be dealt with on an item-by-item basis, but some general principles as to the treatment of corporations, financial intermediaries, and nonresident activities will have to be decided upon if the accounts are to be comparable from region to region and even from item to item.

## *III. THE FUNCTION OF ECONOMIC ACCOUNTS AT THE NATIONAL LEVEL*

Formal economic accounts even at the national level are a fairly recent development. As is the case with most ideas, the history of these accounts can be traced to many different elements in the past, but it is only in the past few decades that they have emerged as a systematic presentation of data. Even today the speed with which developments are taking place in national accounting leads to a feeling of chaos rather than order. At present, at least five different forms of national economic accounts are recognized. These are: (1) national income accounts; (2) input-output accounts; (3) flow of funds accounts; (4) balance of payments accounts; and (5) national balance sheets. A brief description of the nature and function of each of these accounts will be useful before discussing whether regional breakdowns of such accounts are desirable and feasible.

*National Income Accounts*

National income accounts are perhaps the most familiar form of national economic accounts. They are concerned with the income and expenditure flows among producers, consumers, and government. In addition, these accounts show the relationship of the domestic economy to other countries and the amount of saving and investment taking place.

The function of the national income accounts is to help the economist analyze the changes in the level of economic activity taking place in the economic system. The gross national product account shows, on the one side, the level of expenditures by individuals for consumer goods, by governments, by producers on capital account (including inventory change), and by importers and exporters. A change in the spending habits of any one of these groups will have repercussions upon the level of economic activity in the nation, and unless it is offset by an opposite change elsewhere in the economy, it will have a cumulative effect upon the system. Thus, for example, an increase in defense spending by the government, such as occurred at the time of the Korean boom, may have a highly stimulating effect upon the economy and lead to a demand inflation. Similarly, with respect to the capital expenditures of producers, the emergence of excess capacity or a decline in profits may cause a decrease in capital goods expenditures. This may signal a recession or downturn in the level of economic activity.

Changes in the level of expenditures may result, on the income side of the gross national product account, in changes in government tax receipts, profits and similar payments, and in wage and salary payments to individuals.

Given the tax laws and government obligations to pay unemployment insurance, it is obvious that the government account will undergo substantial change with changes in the level of economic activity. A decline in the total expenditures and income in the economy will, on the one hand, cause tax revenues at the existing tax rates to shrink, and on the other hand it will increase government outlays due to the burden of unemployment compensation. As a consequence, a government deficit will be created by a recession and, similarly, prosperity should normally bring with it a government surplus. The national income accounts, thus, are useful as a framework for analysis of the interdependence and interaction of producers, consumers, and government in the economy, and for providing some of the material required for making short-term projections of economic activity.

### *Input-Output Accounts*

The national income accounts do not provide information on the interrelationship among industries in an economy. For a great many purposes, however, such information is important. For example, in economic development plans it is necessary to be able to estimate what the requirements for electric power, steel, aluminum, or other raw materials may be at some future date. For national defense purposes it is necessary to estimate what can be produced with the kinds of basic resources available to an economy. The national income accounts deal only in the aggregates of production and consumption, and such interindustry detail is not available.

In answer to the need for more information concerning the industrial structure of the economy, input-output systems have been developed. An input-output table, by showing the purchases of industries from each other, shows what items of input are required for any given industry to operate and how the output of this industry is distributed among other industries and final users. Countries involved in development plans have been making an increasing use of input-output analysis in order to ensure planned growth In the absence of some such study, it is not possible to forecast the demand for the output of specific industries and thus investment decisions may not be optimal in terms of meeting future requirements.

Recently the importance of input-output relations has also been recognized in connection with the development of price indexes suitable for deflating economic activity by industrial sectors. In any given industry, production can be measured by the value which that industry adds to the total productive activity of the economy. In order to eliminate price changes from the measurement of value added in real terms, it is necessary to deflate the inputs into the industry and the outputs of the industry separately, obtaining as a result a deflated value added. The analysis of projected changes in various industries depends upon our ability to make adequate measurements of output in real terms. Thus, a better understanding of productivity growth and technological change in the economy, as it is related to the production taking place in different industries, will in the last analysis require input-output information.

### *Flow of Funds Accounting*

Flow of funds accounting is a more recent development than either national income or input-output accounting. Its purpose is more closely related to that of the national income accounts than it is to input-output analysis. For many kinds of economic analysis the national income accounts are deficient in that they do not show financial transactions as well as income and product transactions. For example, national income accounts give very little information on the sources and uses of funds for investment. There can be little doubt that the increasing importance of insurance companies and other financial institutions has had considerable impact upon the availability of funds for investment purposes in different parts of the economy. The flow of funds accounts reveal the part played by the financial institutions by showing their purchases of securities and the loans they make to businesses; and the sources of these extensions of credit in terms of the premiums paid for pension plans, life insurance, etc. At the present time our financial institutions are changing rapidly, and these changes cannot be evaluated adequately in the context of national income accounts alone. For instance, if the government wishes to float new bonds it must be able to estimate how much can be absorbed by the various kinds of financial institutions, and the resulting changes in the money supply must be considered in terms of their effects upon the supply of credit in the economy in terms of all forms of financing institutions.

Flow of funds accounts employ many more sectors than do national income accounts. These sectors are not defined on an industrial basis, as in input-output accounting, but rather deal with legal forms of organization such as corporations, unincorporated enterprises, and various forms of financial institutions. The flow of funds accounts are much less consolidated than the national income accounts, and they cover changes in assets and liabilities as well as income and product transactions. Such aspects of government activity as debt management and pension and trust funds are treated in considerable detail.

### *Balance of Payments Accounts*

Balance of payments accounts show the payments and receipts flowing between one economy and other economies with which it has commercial relationships. When a country buys more from abroad than it sells, this trade deficit must be made up by capital flows or actual shipments of gold or by international transfers. The balance of payments accounts spell out these various transactions and show their interrelation. When broken down by country and commodity, the structure of foreign trade and its change over time can be analyzed. Such analysis is helpful in studying what future adjustments a country may be forced to make in light of the changes occurring in its trade patterns. In effect, the balance of payments accounts provide greater detail than national income accounts in the field of foreign trade and payments. They are highly related to input-output accounts, insofar as they pertain to the industrial structure of a country's production. They are also directly related to the flow of funds accounts, insofar as they affect the credit structure and supply of gold and holdings of foreign securities.

### *National Balance Sheets*

National balance sheets are as yet comparatively rare. There are signs, however, that in the near future many countries will be developing these accounts on a systematic basis. The inflationary pressures of the immediate postwar period would have been more readily understood had information been available on the balance sheets of individuals and businesses. At that time, both individuals and businesses held large amounts of liquid assets in the form of cash bank deposits and government bonds. A knowledge of the magnitudes involved would have made economists realize that the most important problem of the immediate postwar period would not be one of recession, as many predicted, but rather one of controlling demand to prevent serious inflation.

Government tax policy, control over the supply of money, and general debt management all affect the assets and liabilities of individuals and businesses, so that both their current and their future behavior is altered. Combatting a recession, for example, by building up the supply of money may create liquidity which will render other government controls impotent in some future inflationary situation. It is gradually becoming obvious that we must pay attention not only to the income transactions and the financial transactions but also to the stock of assets and liabilities held by individuals and businesses.

The national balance sheet is really an extension of the flow of funds accounts. Its sectors naturally must follow legal forms of organization since only legal entities can have assets, liabilities, and equities. Since the flow of funds accounts show changes in assets, liabilities, and equities, the same information would suffice to keep national balance sheets up to date from year to year.

## *IV. THE APPLICABILITY OF NATIONAL ECONOMIC ACCOUNTS TO REGIONS*

It would be a relatively simple matter to suggest that the framework of economic information developed for the nation should be developed also for each of the regional levels. In some degree there would be merit in such an approach—it would permit a better understanding of the process of aggregation and the role which each of the various regions plays in the over-all national economy. However, it is not immediately obvious that such a statistical framework would be feasible, given our present techniques of measurement, nor would it be needed for most uses of regional economic analysis. The following pages will take a closer look at the details of the national economic accounts and consider just what parts of these accounts can and should be regionalized.

## *National Income Accounts*

There have been suggestions that the national income accounts, as a whole, should be built up at the regional level. The rationale for this approach has never been too clear. At the national level the major use of a complete national income accounting system is to study the general level of economic activity and the interaction among the various economic sectors. This type of analysis can be done most easily in the case of a closed economy. As the economy in question becomes more and more open, other forms of economic accounting will be required for the study of the impact which related regions may have upon this region. Thus, for example, the industrial breakdown of a region becomes somewhat more important than such things as saving, investment, and even government expenditure within the region. Put in other terms, it is clear that the major regional problems are different from those posed at the national level, and it does not necessarily follow that they require a complete set of national income and product accounts as their basic framework of economic information.

On the other hand, there are pieces of the national income and product accounts which would be very useful to have on a regional basis. It will, therefore, be necessary to examine the different parts of the national income and product accounts to evaluate which of these can be usefully employed in regional analysis, and which are statistically feasible to produce.

*GNP Expenditure Breakdown.* The gross national product expenditure breakdown consists of five major flows: consumer expenditures, government expenditures, gross capital formation, exports, and imports. The sources of this product information are quite varied and each item will have to be examined separately.

It would be very useful to have data on consumer expenditures on a regional basis. For one thing, a comparison of per capita consumption expenditures in various regions on various kinds of goods would yield considerable insight into the welfare levels enjoyed by various regions, and changes in consumption over time would also be a good indicator of the progress which a given region was making. There would also be a considerable demand for information on regional consumer expenditures by groups interested in establishing retail outlets or production facilities for goods consumed locally. A number of large companies do extensive market surveys within regions to obtain precisely this sort of information; it would be useful if such figures were available to all in the community on a comparable and reliable basis.

Unfortunately, consumer expenditures by region are difficult to estimate. In the national accounts of the United States the basic method employed has been the establishment of primary benchmarks by the commodity flow method, using the Census of Manufactures production data and adding on wholesale and retail markups as the goods flow into channels of distribution and are purchased by consumers. Services pose other kinds of problems. Reliance in this case is placed upon the Census of Business and tabulations of income from the professions obtained from the Internal Revenue Service. These techniques of making estimates of consumer expenditures will not at any juncture yield regional details. It is quite possible, for example, to know how much of a consumer good is produced, marketed in the United States, and exported, but it is quite a different thing to know how much each region of the country purchases of this commodity. If we are to get data on consumer expenditures by region, it will be necessary to develop new statistical sources and use new techniques. Offhand, there does not seem to be any general, across-the-board kind of statistical information which is available for all regions and which would yield satisfactory information on consumer expenditures. There are, therefore, two possible procedures. One is to institute a large-scale sampling technique which would yield regional consumer expenditure information; the other is to leave the problem of consumer expenditures to each region individually. With respect to the first alternative, there might be some hope in making use of the sample budget studies required for cost-of-living price indexes. However, the sample surveys do not cover cities below a certain size, or rural communities. As a result, it would be hard to build these sample surveys up into estimates which could serve for particular regions. In view of the costs involved, therefore, it does not seem feasible at this juncture to provide consumer expenditure information on a centralized basis for all regions. For the moment, this suggests that the task should be left mainly to those individual regions that would find it worth while to build such information up from existing fragments of information that are available, plus any additional sampling that might be required.

Government expenditures on goods and services are also desirable in order to evaluate the impact such spending has upon particular regions. This is especially important where a region may be a major center for a defense facility or for the production of defense material. Any appraisal of the possible impact of disarmament, or of an increase in the pace of the cold war, will involve differential regional impacts. It is also important to have information on state and local expenditures in a region in order to evaluate their welfare impact, in a manner similar to that already discussed for consumer expenditures. Thus, for example, data on expenditures on education per capita in different regions are useful for both comparative purposes and analysis over time.

The materials available for making regional estimates of government expenditures should be somewhat better than those for consumer expenditures. Unfortunately, to date the federal government's outlays have not been classified in a manner to make possible regional computations of expenditures on goods and services. Some information is available on disbursements, but in large measure it covers more than merely the expenditures on goods and services. At the state and local government levels, the Census of Governments should be able to provide the desired kind of information, and there is no particular reason why this should not be done centrally.

Gross capital expenditures by region would be very pertinent to the study of regional growth and changes in productivity.

For both residential and nonresidential construction, the present method of collecting information should be capable of yielding regional data if this were made one of the major objectives of the tabulation. The main question here would be whether, because of considerations of reliability, data for many regions would be sketchy and unsatisfactory. Nevertheless, it would seem that with some additional input of resources satisfactory regional data on residential and nonresidential construction could be maintained. With respect to producers' durable equipment, the traditional approach here has also been the commodity flow method. Data contained in the Annual Survey of Manufactures and the plant and equipment series of the Securities and Exchange Commission should be sufficient to provide some sort of regional breakdown which would be useful. The importance of producers' durable equipment on a regional basis is so great, and the raw materials at hand so much better than in the case of the consumer expenditures, that additional efforts on a centralized basis would seem to be justified in order to provide this information at the regional level.

Data on imports and exports at the regional level would be very useful if the imports and exports were defined with respect to the region in question, rather than at the national level. However, this information is completely nonexistent and there are no obvious sources which can be drawn on to produce it centrally. At some later date, if it is possible to obtain complete production and expenditure data for each region, it might be possible to obtain the net imports and net exports by commodity group, but, even then, derivation on a gross basis would remain a problem. For the time being, important as such data are for regional analysis, it would seem that they must be left for special studies on a decentralized basis.

*GNP Distributive Shares.* In obtaining information on the gross national product by distributive shares within a region, it is necessary to differentiate between two concepts: first, income that is produced in the region and retained in the region; and secondly, income produced within the region that is paid to other regions. Thus, data on wage and salary disbursements must be divided between those paid to residents in the area and those paid to residents of other areas. As has already been mentioned, at the state level where large metropolitan areas extend into several states, this problem may be somewhat difficult. Where there is not much commutation from outside the region, however, the disbursements of wages and salaries should approximate the receipts of wages and salaries by residents of the region, and the problem will be minimized.

The major problem in estimating distributive shares arises with multi-plant corporations, with respect to the allocation of dividends and undistributed profits. From a legal point of view, the corporation is located in the region where it has been incorporated. This would suggest that New Jersey and Delaware were large payers of dividends to the rest of the country, and large receivers of corporate profits and payers of corporate profit taxes. If this fiction is abandoned, however, some decision must be made as to the regional allocation of dividend disbursements. Similarly, the overhead of corporate head offices, corporate profits, corporate profit taxes, and undistributed profits must be divided among the areas in which the individual producing units of the corporation are located.

Some solution must be adopted before we can regionalize gross national product by distributive shares. Given the basic form of the statistical information from the Census of Business and the Internal Revenue Service, furthermore, it seems reasonable that these estimates should be made centrally rather than on a decentralized basis. As will be seen in the discussion of the personal income account, the Department of Commerce is already heavily involved in making regional income estimates. For purposes of comparability among regions and in order to ensure that the separate regional estimates add up to a national total, an attempt should be made to sort out regionally the disbursements of income as well as the receipts of income.

*Personal Income Accounts.* At the present time the Department of Commerce publishes personal income by states, regions, and metropolitan areas. These data are used extensively in regional analyses as measurements of the level and change in level of income in the various regions. In many ways the personal income figures are the cornerstone on which the regional accounts are being built, and it may be this concept, rather than the concept of gross national product, which will become most useful for regional economic analysis.

Whether it is possible on a centralized basis to extend the present personal income estimates into estimates by size of income remains to be seen. Certainly, a great deal of the information going into the present personal income estimates would be useful in deriving estimates by size class. On the other hand, major gaps would remain to be filled, and these would have to be filled by sample surveys as is now done nationally by the Current Population Surveys. To obtain satisfactory regional detail might well require more resources than the results would be worth. Certainly, yearly size distributions of income by region are not feasible or practical; it may well be that the estimation of size distribution should be mainly carried out on a decentralized basis for such regions as desire them.

With respect to the expenditure size of personal income, comments have already been made concerning the difficulty of obtaining consumer expenditures. Estimates of the taxes paid by individuals in the various regions are currently being made by the Department of Commerce in the derivation of disposable personal income on a regional basis.

*Government Accounts.* Government accounts on a regional basis need to be broken down into two categories. First, there are the accounts of governments within the region, i.e., state or local governments. Second, there are those of governments outside the region, i.e., federal or state governments. In the case of governments within the region, data are provided in the Census of Governments. The major problem lies in determining the relation of a region to governments outside of the region. This difficulty has already been noted in the discussion of government expenditures.

The same observations that were made with respect to goods and services are pertinent to the estimation of transfer payments made by outside governments to recipients within a region. On the tax side, it is important to know how much taxes individuals and businesses within a region are paying to the various governments. Such information is necessary for studies concerned with the analysis of the extent to which a government is draining funds from or injecting funds into a region. Again, for reasons of comparability between regions and because of the desirability of having all regional figures tie into official totals, it is highly desirable that such estimation be carried out on a centralized basis.

*Imports and Exports.* As has already been noted in the discussion of expenditures for gross national product, it does not seem reasonable at this juncture to make an over-all effort to trace the interregional trade flows. The most that could be done would be to make regional breakdowns of production in as detailed industrial groupings as are possible without disclosure, as one of the raw materials in the estimation process.

### *Input-Output*

Input-output tables have been drawn up for several regions and have proved to be useful in setting forth the structure of a region so that its sensitivity to cyclical behavior can be better understood. They are also useful in analyzing the impact on a regional economy of changes in national trends and technological changes. There is little hope at this time of systematical production of regional input-output tables on a centralized basis. Nevertheless, as in the case of imports and exports, provision of basic information about production taking place in the various regions greatly facilitates the task. It is probably true that it would be somewhat wasteful of resources to produce input-output tables for all regions. In any case, as is true with national input-output tables, they would not be required on a yearly basis.

### *Flow of Funds*

Flow of funds accounts at the regional level run into many of the same difficulties as have been discussed in connection with the gross national product by distributive shares. On the one hand, if strict legal definitions are followed it would appear that particular regions of the economy are dominant when in fact these regions merely reflect a large burden of gross transfers. Thus, for example, the home office of a life insurance company might be credited with the receipts and disbursements of a large amount of funds despite the fact that in reality these transactions would be taking place in other regions.

Nevertheless, it would be extremely useful to have flow of funds information on many groups within a region. Thus, for example, the changes in assets and liabilities for consumers, which would reflect such things as the changes taking place in consumer bank deposits and changes in the stock of durable goods which they hold, would be extremely useful in analyzing potential changes in demand within the region. Similarly, an analysis of the financial sector within a region would be useful in explaining behavior of the construction industry and inventory accumulation.

At this point in time there seems to be little virtue in building up a comprehensive interrelated flow of funds system for the various regions of the nation. Nevertheless, within the specific regions there are probably certain groups for which flow of funds material should be built up. These would include consumers and specific kinds of financial institutions.

### *Balance of Payments*

Information on the balance of payments between regions would be very useful in showing which regions are traditionally in debt to other regions. It is through balance of payments studies that the process of growth and differences in rate of growth can be understood. A considerable amount of work has been done on the interrelation of the different Federal Reserve Districts, but it has been quite difficult to carry balance of payments data below this level. There are obvious advantages in carrying out the balance of payments calculations on a centralized basis, since the very nature of the estimation process gives one a cross-check as to the consistency of the estimates. Nevertheless, the task of developing meaningful to-whom–from-whom balances is so great that continued experimentation at the individual-region level seems to be in order.

### *Wealth and Balance Sheets*

Wealth and balance sheets at the national level are just beginning to be developed; it is probably somewhat premature, therefore, to suggest that we should try to build up a set of interregional balance sheets even before we have official ones for the nation as a whole. Nevertheless, the provision of information on assets and liabilities of individuals and producing units within a region would do much to enhance our knowledge of welfare, productivity, and growth. Because the real wealth of a nation must, in fact, be located geographically, it is not impossible to build the wealth estimates up on a regional basis. The Census of Governments tells us much about real estate assessments and even about the value of plant and equipment of business. Much of this material can be developed centrally and provided on a regional basis without too many problems of disclosure.

### *Conclusion*

The examination of national accounts on an item-by-item basis strongly suggests that straightforward disaggregation of the complete accounts is not the solution. There are too many places in the national accounts where the introduction of a regional breakdown is not feasible due to the procedures by which the national estimates are made. Thus, for example, data on consumer expenditures cannot be provided out of the same body of information that is now used to derive the national totals. In other cases, regional data require a new kind of information not in fact contained in the national accounts. Thus, for example, interregional trade flows and the balance of payments between regions, which are consolidated out of the national figures, are of great importance in the regional data.

Any thoroughgoing research at the regional level will probably require the kind of work carried out by Werner Hirsch[1] and Charles Leven,[2] involving special data collection at the regional level. A simple disaggregation of the national economic accounts might seriously limit the analytical usefulness of the regional accounts. The economic accounts required at the regional level will be oriented quite differently from those at the national level, and the form of the regional accounts should be adapted to the uses to which these accounts will be put.

This line of argument does not mean to suggest that the construction of regional accounts and the construction of national accounts are independent undertakings. In the first place, it will often be found necessary to combine regional economic analysis with economic analysis at the national level. If we desire, for example, to examine the impact of something taking place in the nation as a whole upon different regions we will require a set of national accounts which can be regionally articulated. The kind of detail here, however, may be quite different from that required for the detailed analysis of what may be purely regional economic problems. Thus, for example, if one wishes to examine the role of the local government in a local economy, the kind of regional data provided by the breakdown of national data might be wholly inadequate.

Another reason why national accounts need to be disaggregated, at least in part, to the regional level is to ensure comparability between the different regions being analyzed. This is especially important since, as has already been noted, one of the major tools of regional economic analysis is the comparative study of the levels at which different regions are operating and the behavior of the regions over time. The advantage of a disaggregation of national totals is that it ensures that the same types of statistical sources and the same methodological approach are used in developing the estimates so that differences in data from region to region arising from these elements can be minimized.

[1] *The Review of Economics and Statistics,* 41 (November 1959) 360-69.

[2] See "Regional Income and Product Accounts: Construction and Applications" in this volume, pp. 148-95.

Some discipline is also required to make sure that the regional accounts, which are being constructed by different research workers in different areas, are not so dissimilar that regional work becomes fragmented. If a large number of groups are working along interrelated lines, improvements by one group can be applied in similar circumstances by other groups working in other regions. To this extent, then, some common framework, at least in broad outline, would be useful as a major point of departure for regional economic research.

From a computational point of view, it is important that the data in the national economic accounts be disaggregated on a regional basis wherever this is feasible in terms of the basic statistics involved. It is recognized that the regional figures may not possess the same degree of reliability as the national figures; nevertheless, the experience with the personal income data by region has indicated the value of such efforts. There are, of course, problems of disclosure that limit the availability of the Census of Manufactures and Census of Business data on a plant-by-plant basis, or even for smaller communities on a 4-digit basis. Nevertheless, it would be extremely helpful if such data could be made available in a form which did not result in disclosures. Thus, for example, the value added by even broad categories of manufacturing (2- or 3-digit level) would be a considerable help to those building regional accounts, reducing their data collection problem and increasing the reliability of their work.

It would also be useful if an attempt were made to produce additional economic constructs at the regional level. At present, those of personal income and disposable income exist, but it would be helpful if constructs of gross national product and some of the major flows such as tax payments, investment expenditures, etc., could be produced on a regional basis. It is only by introducing such constructs on a centralized basis that common points of departure can be set up for regional economic accounts.

# COMMENT

*Werner Hochwald*

*Washington University*

The Conference on Regional Accounts is fortunate to have authorities like the Ruggleses express their interest as affirmatively and persuasively as they do. Their discussion is concerned with two basic questions: (1) Are economic accounts a useful approach to regional analysis? and (2) Are the national accounts a proper guide to the construction of regional accounts?

The first question can be answered in the affirmative, I think, without much hesitation, as has been implied in the Ruggles paper and the very theme of this Conference. Though it is true that no real consistency check is forthcoming unless the accounts are built up from substantially complete sets of measures, even a modest scale of effort should gain from the internal discipline inherent in any double-entry bookkeeping system.

More difficult is an answer to the second question: Are national accounts a proper guide to the construction of regional accounts? It may be helpful to restate the issues somewhat. Rather than emphasize the essential differences between national and regional accounts, we could visualize the "region" as a new system of aggregating individual components in the national accounts, either as they now exist or as they should be created. Looking at the problem this way, the question turns into an examination of the components and sectors most useful for regional as well as national analysis.

The choice of components for regional aggregates suggests two basic alternatives in the definition of a "region." Components for any system of accounts may either be "natural" transactors, such as individual persons, or "legal" transactors, such as corporate entities. And these transactors, in turn, may either be related through physical transactions, such as the actual work done in the shop for a customer, or they may be related through legal relations, such as the incurrence and discharge of debt. The subtle interactions of physical and legal phenomena have been the continuous concern of economists in their attempts to penetrate the veil of institutional and monetary arrangements toward an understanding of the "real" world. In the national accounts we permissibly slur over some of these difficulties as the geographic entity and the institutional entity of the nation closely overlap. In the regional accounts, the distinction becomes of crucial importance as residents have a wide divergence of property and debt relations with the rest of the country. In fact, the very challenge of regional development in an open economy is the high rate of substitution among highly mobile national factors in their shifting relations with less mobile regional resources.

Regional breakdowns of national economic accounts thus depend on the type of region we wish to describe. The geographic region, held together by immobility of natural resources and physical activities to transform these resources into movable consumer goods, is described most readily by the disaggregation of national accounts composed of personal income recipients or physical establishments related to each other through commodity flows, such as the cells of a national input-output table. The political or institutional region, held together by the common property interests of its residents related to each other and "foreigners" through the national debt structure, is described by a regional balance of payments, where the region becomes a sector in the statement of national money flows. Once we accept the region, however defined, as a sector in the national accounts, the task of constructing regional accounts is partly a breakdown of national totals, wherever feasible, and partly a build-up of new national accounts composed of regional aggregates. The Ruggles paper has shown the limited usefulness of national economic accounts, as we know them, for regional breakdowns. The major task therefore remains to build regional accounts from the ground up to form a new national system for the recording of spatial immobilities in an open economy.

The magnitude of this task calls for carefully considered priorities to economize scarce resources of economic research. Two possibilities stand out because of their obvious promise for regional as well as national analysis. First, there appears wide agreement about the desirability of recording gross national product by industrial origin as an essential step toward the accounting of gross regional product. Second, there is the even greater challenge of implementing the interaction model as a first step toward true double-entry economic accounting where each flow is simultaneously recorded as a change in assets. It is this need of regional analysis for the accounting of comparative spatial advantage which may yet bring the most striking changes in our regional as well as national accounts.

## *Charles F. Schwartz*

*International Monetary Fund*

My chief disagreement with the paper by Richard and Nancy Ruggles is that it does not hit hard enough at the infeasibility of estimating some components of the national economic accounts on a geographic basis. However, there is room for differences of judgment; and I would surmise from their paper that the Ruggleses agree in substance with my own views that (1) presently available data would be inadequate for the purpose of comprehensively extending the national accounts to an area basis, (2) the collection of additional requisite statistics would, in important instances, be difficult and costly, and (3) there does not appear in any event to be an impelling, demonstrated analytical need to duplicate for subareas the accounting system which has been found appropriate for the nation as a whole.

On a state basis, most needs of analysis have been met by the existing estimates of "personal income." This is a very broad economic measure; it can be estimated rather reliably on a geographic basis; and it is susceptible to industrial breakdowns which, as emphasized by the Ruggleses, are especially valuable for area analysis in the United States.

In the regional phase of national income accounting in Washington, much still remains to be done as a logical extension of prior work. (a) Estimates of personal income on a regular annual basis (with some back-year reference points) should be prepared by the government for each metropolitan area in the country. Such estimates would be quite feasible statistically—especially with the recent decision by the Internal Revenue Service to tabulate for metropolitan areas numerous items reported on federal income tax returns—and would be extremely valuable to business firms and to economists and administrators grappling with the diverse problems arising from the rapid growth of urban centers. (b) The estimates of personal income should be extended to counties as well as metropolitan areas. Here, however, the basic work should be done by some local agency in each state, such as a university bureau of business research or a state government department. It would not be practicable for the Commerce Department's regional income staff in Washington to attempt to estimate personal income for the more than 3,000 counties, but it could do a great deal to guide and co-ordinate the work, stimulate the flow of basic data, and see that uniform statistical sources and procedures were used throughout the country. The best administrative arrangement for this furthering of county income measurement would be co-operative programs between the federal government and the individual states. (I am aware, of course, that the estimates for many counties would be weak, perhaps unusably so individually, but one of the main purposes of county income work is to furnish the building blocks for larger, more economically significant groupings.)

Such a vast body of data—personal income by states, metropolitan areas, and counties—would provide a comprehensive, yet detailed, foundation of regional economic and market analysis. Conceptual and statistical comparability would be a major feature of these data, so that any area could be studied in relation to any other and to the nation with respect to level of income, industrial structure, and patterns of economic change. Furthermore, the framework provided by personal income could be broadened in systematic fashion to cover employment, population, and other meaningful types of statistics. It could be supplemented at the local level by special-purpose projects as research and experimentation in regional income accounting continued to go forward outside of Washington along many different lines.

6

# REGIONAL INCOME AND PRODUCT ACCOUNTS: CONSTRUCTION AND APPLICATIONS

*Charles L. Leven*[1]

*University of Pennsylvania*

FUNDAMENTALLY, A SET of social accounts is an empirical framework corresponding to a theoretical structure which postulates the nature of relationships between various aspects of some particular phenomenon. Sometimes the structural relationships are referred to as "accounts," i.e., the accounting framework itself. In other circumstances the term is meant to refer to the set of values set forth within the framework. For the sake of aiding the exposition, "accounts" will be used here to cover both meanings; the distinction between the two uses should be clear from the context.

The application of social accounting techniques to any particular

[1] The author wishes to express his gratitude for the generous assistance provided by the Sioux City Planning and Zoning Commission, especially W. Burnett Austin and Arlo Herbold, and for the financial assistance of the Social Science Research Council. For helpful comments on the manuscript the author is indebted to Professor Charles M. Tiebout, of the University of California (Los Angeles); Professor Werner Z. Hirsch, of Washington University; Professor Benjamin Stevens, of the University of Pennsylvania; and especially to his students in the Department of Regional Science at the University of Pennsylvania.

phenomenon depends, however, on the validity of three propositions: (1) that the phenomenon is resolvable into a discrete number of components; (2) that the intensity of the phenomenon is the weighted sum of the intensities of its respective components; and (3) that the intensity of each of some subset of the components is dependent on the intensities of all of the other components and of the phenomenon itself, while the intensity of each of the other components is dependent on other phenomena.[2] Where these conditions hold, the accounts that are constructed can serve important analytical uses. First, they provide a basis for analyzing the effects on the phenomenon of changes in the independently determined components. Second, they enable a comparison of the relative importance of each of the components. Third, they can afford a theoretical perspective for making predictions about the phenomenon, both by way of isolating the critical elements of change and by providing a framework for evaluating the consistency of the prediction, at least with respect to known phenomenon-component relationships.

Moreover, it could be argued that both the second and third uses are methodologically subsidiary to the first. What it is that we want to evaluate the relative importance of, or to predict, can be meaningfully identified only after relevant phenomena have been specified and the relationships between these phenomena and their components analyzed.[3] Such an argument depends, admittedly, on a conception of accounts analysis as necessarily stemming out of an interest in some preconceived question, be it the testing of a theoretical hypothesis or a "practical" policy issue. Insofar as accounts represent an empirical application of some theoretical formulation, such a conception would seem appropriate. By analogy, while the motivations of business behavior might be investigated without any preconceptions as to the findings or their significance, to make up a set of books for a corporation without any

[2] While the validity of only the first two propositions would permit the formulation of accounts it should be clear, at least from subsequent discussion, that such accounts would be of no analytical significance.

[3] This is meant to apply only to welfare or predictive analyses carried out within an accounts framework. Such analyses, of course, could be undertaken by means of some entirely different approach.

idea of what they are trying to maximize would seem a rather pointless pursuit.

The preceding remarks have not been phrased in such tortured generalities without reason. They are intended to indicate clearly and explicitly that throughout this paper accounts will be conceived of exclusively as a tool to be used in the analysis of particular questions. This is meant to stand in opposition to a conception of accounts, especially some particular accounting framework, as having an existence or significance of their own. Perhaps this discussion could have been avoided merely by stressing the importance of asking the right questions. However, in view of the kind of resistance to change which apparently has been built up in the national income and product accounts (with respect to the treatment of durable and nondurable consumers' goods, for example), it seems wise to emphasize the need for flexibility while regional accounts are still in their youth. Moreover, such flexibility need not be at the expense of internal consistency between different sets of accounts for a given area or the reconcilability of regional and national accounts. But more will be said of this later.

To bring the discussion down to a less abstract level, we can begin by substituting "activity" for "phenomenon" and "sector" for "component." This more or less focuses our attention on economic phenomena and enables the accounts framework to be written symbolically,

$$(1) \qquad A \equiv \sum_{i=1}^{m} A_i + \sum_{j=m+1}^{n} A_j$$

$$(2) \qquad A_i = f_i(A_1, \ldots, A_n, A) \qquad (i = 1, \ldots m),$$

where $A$ is the level of the activity, $A_i$ are the endogenous, and $A_j$ are the exogenous sectors. The system then consists of one account identity and $m$ equations in $(n+1)$ unknowns. These $(m+1)$ relationships can be solved for $(m+1)$ of the unknowns, ordinarily $A$ and $A_i$, in terms of the other $(n-m)$ variables, giving the system in reduced form,

$$(3) \qquad A = \sum_{i=1}^{m} A_i + \sum_{j=m+1}^{n} \sum A_j$$

$$(4) \qquad A_i = g_i(A_{m+1}, \ldots, A_n) \qquad (i = 1, \ldots m).$$

But this is still a rather abstract conception and if an accounts framework is to be constructed relative to some particular analytical problem, it is necessary to identify and specify the dimension of $A$, the relevant phenomenon, to postulate its components, and to partition this set of components into two subsets, $A_i$ and $A_j$, the endogenous and exogenous sectors, respectively. The national income accounts, for example, are constructed on the basis of only one particular specification and set of postulates. Obviously many other specifications would be possible, and some of these well might be useful in the context of particular policy issues. In the material to follow an attempt will be made to investigate the kinds of account systems that emerge from considering a few of the issues of importance at the regional or metropolitan level—with the analysis always proceeding from an identification of the problem to the necessary specifications for implementing the generalized accounts structure to working out the details of the accounting framework. This is meant as an alternative to simply seeking out the kinds of policy issues and the extent to which such issues can be analyzed within a preconceived accounting framework.

Perhaps the phenomenon of greatest interest to economists is an area's income and product, along with the corresponding employment of labor resources. Thus, in Section I evolution of accounts appropriate to analysis of this phenomenon will be discussed in detail. Since this is the same objective as that of the national income and product accounts, it should not be surprising that the results will bear a considerable resemblance to those of the Department of Commerce; but there will also be significant differences. One of these differences concerns the critical role of interindustry relationships in regional accounts, which will be covered in Section II. Section III will include some suggestions for modifications or supplements to the accounts structure needed to handle some other analytical objectives. Section IV will consider very briefly some of the problems of empirical implementation of the accounts, with special reference to filling the informational requirements of regional accounts out of Census and national accounts data. Some examples of empirical results will be included in the Appendices.

## I. REGIONAL INCOME AND PRODUCT ACCOUNTS

It will be convenient to consider the income and product accounts for a region as the basic set of accounts, with those stemming from other objectives as modifications of them. While this convention is arbitrary, and while accounts related to any alternative policy question could have been chosen as central to the analytical scheme, the results would not differ in any essential manner, but only in the sequence of analyses for which they were designed.

A starting point in laying down the specifications for regional income and product accounts is to consider the applicability of the national income and product accounts. In their simplest form the underlying theoretical specifications of the latter can be indicated by,

$$Y \equiv C + I + G + (e - m) \tag{5}$$

$$C = a + bY \tag{6}$$

where $Y$ is gross national product, $C$ is consumption expenditure, $I$ is gross private domestic investment, $G$ is government purchases of goods and services, $(e - m)$ is the balance of trade, with $e$ representing exports and $m$, imports. Moreover, the accounts corresponding to this model can be derived from the double definition of $Y$, both as gross national product or *gross* national income, namely,

$$Y \equiv \left[\sum_{i=1}^{N} s_f^i\right] - m \tag{7}$$

$$Y \equiv \sum_{i=1}^{N} v_p^i \tag{8}$$

where $\sum s_f^i$ is sales for final use and $\sum v_p^i$ is value added in production by the $i^{th}$ enterprise, $N$ is the number of enterprises in the economy, and $m$ is imports, as above. Equating the right-hand sides of these two identities gives us the simplest possible set of accounts, namely one with a single T-account as follows:[4]

*Consolidated income and product account*

$$\sum_{i=1}^{N} v_p^i = \left[\sum_{i=1}^{N} s_f^i\right] - m \tag{9}$$

This consolidated account could be disaggregated, of course, into individual sector accounts as indicated by (5), but there would seem to be no need to repeat these results here.

The national accounts appear to adequately isolate the exogenous elements in income determination and to provide a framework for forecasting gross national product on the basis of projections of these exogenous variables. But with regard to analyzing the relative contributions to GNP of various enterprise groups, or industries, certain complications appear which require further modification. More specifically, that subsidiary tables showing GNP by industrial origin do not appear in the national income statistics is not an accidental omission. For one thing, this would require an allocation of capital consumption allowances, which are estimated on an aggregate basis, among the several industries. But this is not the basic problem, since tables showing net national product by industrial origin are also missing. That these latter figures cannot be derived meaningfully stems from two additional difficulties. First, as noted above, net national product includes indirect business taxes. Such taxes vary substantially as a percentage of total revenue from industry to industry, and they are generally regarded as primarily representing a tax on the purchasers of the commodity and only incidentally as compensation for productive services pro-

[4] If the definitions of value added and sales for final use which are used in the Census of Manufactures and the national income accounts are used, the equation will not be satisfied, as these definitions are not entirely consistent. This can be taken care of easily by adding casualty insurance premiums paid by enterprises to value added and subtracting indirect business taxes from sales for final use. Under these adjusted definitions, care must be taken later on *not* to consider indirect business taxes paid to nonlocal governments as an outflow from the area.

vided by government. Consequently, variations in the value of net national product industry by industry reflect in part differences in the extent to which consumers of different commodities are taxed, rather than differences in the net value of product produced. But, as indicated in footnote 4, this problem, at least for a region, can be taken care of quite easily by simply ignoring both the tax component of final sales and such tax payments to nonlocal government. In essence we can abstract from area enterprises' functions as excise tax collection agents.

The other problem in developing figures on net national product by industrial origin is not so easily disposed of, namely, that the net value of product produced in different industries includes varying proportions of imported raw materials. To correct for this bias only domestic contributions to net output are considered, thus leaving only payments to factors of production or national income by industrial origin, as appears in the national income statistics. In analyzing the left-hand side of the income and product account, about the same procedure could be followed for regions as for the nation without running into any serious difficulty. It is with respect to handling the right-hand side of the account and analyzing sectoral interactions that the direct application of national accounts techniques to a metropolitan or regional economy finally breaks down.

This breakdown arises primarily out of the very substantial differences between national and local economies both with respect to the relative importance of imports and the diversity of production. For example, in considering the impact on the national economy of an exogenous change in the level of final demand, the aggregate investment or foreign trade multiplier would ordinarily be applied in a straightforward way. To be sure, depending on the commodity composition of the change in final demand, there would be differences in the magnitude of the total impact. One factor accounting for such differences would be variations in the extent to which intermediate goods were imported from abroad. However, since we are usually working with highly aggregated complexes of commodities and since, on the average, import requirements are a small proportion of total inputs, it seems safe to assume that the change in final demand will occasion the change in imports, indicated by the aggregate marginal propensity to import.

The inappropriateness of following such a procedure for a metropolitan area, or even a larger region, should be readily apparent. For one thing, the variation in import input requirements as between industries in a metropolitan area is enormous, ranging from virtually zero in the case of professional services to 80 or 90 per cent for some manufacturing industries, say meat-packing, and almost 100 per cent for some items (for example, transportation services or power sales contracted for from a local sales office of a firm having no other facilities in the area). Moreover, the average import input requirement is many times that of the United States for even the largest metropolitan areas.[5] Finally, the range of commodities produced in a region is certainly less than for the country as a whole, and in the case of a change in the level of final demand, particularly in the export sector, a rather small number of industries is likely to be involved. What seems most needed for metropolitan accounts is precisely what cannot be derived from the national accounts, namely, either net national product or gross national product by industrial origin. In short, some way must be found to get at imports by sector of final demand. This, then, is the first major difference between regional and national income and product accounts.

Another difference between national and regional accounts, as they are to be developed here, stems from the inappropriateness of the national accounts definitions of exports and imports in dealing with the relationships of a small region to the rest of the world. In the national accounts, factor payments to U.S. residents, either

[5] The New York Metropolitan Regional Study found that even for a 22-county area with a population of 16 million, that about two-thirds of total manufacturing employment was producing goods for export. Even with allowances for their probably having a somewhat favorable balance of trade, higher than average labor productivity, and lower than average propensity to consume, it seems safe to say that probably at least one-third of that area's final demand is satisfied by imported goods, either directly or indirectly. See Max Hall (ed.), *Made in New York* (Cambridge: Harvard University Press, 1959), p. 4.

from or while abroad, are treated as payments for exported services. Similarly, factor payments to nonresidents are regarded as imports on trade account.[6] Such treatment rules out, by definition, any possibility of a difference between the value of product produced in the United States and income accruing (other than by way of transfer payments) to U.S. residents.[7]

For an area as extensive as the United States and having relatively low density of settlement on both sides of its borders, and where capital ownership of and by foreigners is relatively small, such a *tour de force* probably causes little distortion, while considerably simplifying the accounts framework. For a metropolitan area such a convention clearly would be out of the question. The issues with respect to labor commutation, nonresident ownership, and net indebtedness to the rest of their world are highly significant ones and must be considered explicitly. Accordingly, in the regional accounts to be developed below, exports and imports will be redefined so as to include flows of all goods, but only of services other than those of factors of production.[8] The balance of trade will include, of course, exports and imports of both "visible" and "invisible" goods and services. It will not include unilateral transfers.

Finally, the third structural distinction between national and regional accounts will involve a different treatment of government. A region, unlike a nation, may contain within its borders establishments of its local government(s) and also establishments of nonlocal governments.

A significant characteristic of federal establishments within a particular area is that their activity levels are determined predomi-

[6] Unilateral payments have been alternatively treated as either part of the balance of trade or not taken into account at all in the national income and product accounts. In any event, they have not been treated as a current flow on other than trade account.

[7] Actually the value of product produced is forced to equal the value of factor and government income.

[8] It should be noted that payments to self-employed professionals from abroad will still be regarded as arising out of the export of services, but no receipts from abroad of wages and salaries, interest, rent, dividends, or net income from proprietary ownership in nonresident enterprises will be so regarded.

nantly by forces entirely outside the region. In essence, they are producing services for the United States as a whole. Consequently, such establishments will be treated as part of the Rest of the World sector and sales to them by area establishments will be regarded as export sales. Resident employees of such establishments will be regarded as unincorporated enterprises purveying labor services on a contract basis. Wage disbursements by federal establishments in the area to them will be regarded as unincorporated enterprise income from abroad, all of which the recipients pay to themselves as wages. This rather complicated arrangement is merely to avoid treating such employees as out-commuters, and hence biasing the physical dimensions of labor commutation, a magnitude with important implications for physical planning. An exception to this general rule might well be made in the case of the Post Office, and certainly for federal government enterprises. Such establishments could be treated as if they were private nonprofit corporations, with any subsidization of them treated as either a transfer payment or export receipts, depending on the circumstances. Certain other federal activities, such as federal courts and law enforcement agencies, and certain regulatory organizations, obviously do vary in their level of activity according to the level of economic activity in the area. However, establishments in this category usually are of little enough significance, quantitatively, to render special treatment unnecessary. Finally, even the most generalized kinds of federal services—national defense, for example—are partly "consumed" by the residents of the area producing them. These, too, probably are not quantitatively significant except for areas with a sizable share of U. S. population. Handling of such cases will be covered in the following paragraph.

State government establishments in a metropolitan or other substate area can be handled in exactly the same way as federal establishments, with one important exception, namely, where the area contains an important share of the state's total population.[9] In such cases, even the generalized services of state government would importantly serve to satisfy the needs of the area. The sim-

[9] Whether or not the state capital were in the area would be irrelevant.

plest way of handling this situation would be to continue regarding area establishments of state government as part of the Rest of the World, but to regard that share of their wage bill occasioned by servicing the area residents as payment for imports from abroad, for final use by households.[10]

Local government enterprises financed by user charges will be considered as private nonprofit business establishments. Other local government activities (including any subsidy to or surplus of local government enterprises) will be treated as households.[11, 12] Their current outlays will be regarded as personal consumption expenditure and their capital outlays as gross investment. With respect to current outlays, this procedure is justified by the relatively small share of area income arising out of local government employment and the small share of local government services which might be construed as intermediate business product (a very large share of current outlays is accounted for just by public education which is a service to area residents almost exclusively in their capacity as citizens, regardless of how they are paying for it). Since local governments typically employ capital budgeting, and since their definition of capital formation is ordinarily quite close to that used in the national accounts for business investment, it seems pointless not to take advantage of this distinction. Since current outlays of local government have been lumped with personal consumption, their capital outlays will be incorporated into the investment sector simply for convenience. They easily can and should be retained as a separate component of investment in the detailed supporting tables.

A consequence of amalgamating local government into the private sectors is that while all of their product transactions (purchase and sale of goods and services, including labor) are preserved, their

[10] Note that imported government services are valued at factor cost and not incident taxes.

[11] Local government ordinarily would include county government. Even for a metropolitan area, if it is an SMA it will contain an integral number of whole counties, and, where an "urbanized area" is used, the preponderant share of county population would ordinarily be located within it.

[12] That this treatment of local government makes municipal employees analytically equivalent to domestic servants is only a semantic accident.

financial transactions (tax receipts and transfer payments including interest) internal to the area are suppressed. Their financial transactions with the Rest of the World, of course, will be preserved. For the analysis of income, output, and employment in metropolitan areas the implied treatment of domestic financial transactions of local government as interpersonal transfers seems workable, particularly in light of the rather stringent constitutional limitations under which most of them conduct their financial affairs. For a larger region these internal transactions might have to be "unsuppressed." Also, in any situation these conventions would be inappropriate for a social accounts analysis of local government itself. Modifications in the basic set of accounts necessary for this specific application will be discussed in Section III.[13]

A somewhat peculiar convention will be followed for financial institutions. In their product transactions they will be regarded as business enterprises, but in their financial dealings they will be considered as households. This means that, like local governments, their domestic financial transactions will be suppressed but their financial transactions with the Rest of the World will be preserved. This procedure was followed in the empirical studies included in the Appendices largely for strategic reasons, although it can be defended to a considerable extent on the basis of the much greater importance of the national than the local credit market, even for many large metropolitan areas. Basically, it is a way of avoiding some informational problems at what appears to be a not very serious analytical cost. Where research resources permit, it could be avoided.

Private nonprofit institutions will be included in households, as in the national income accounts. Gross domestic investment (including local government) will be defined in the same way as in the national accounts with one exception, namely, that changes in inventories will be ignored (actually they will wind up in the

[13] The suppression of all of government into the other three sectors is a characteristic of the accounts developed here, but not a critical one. All of the government sectors could be made explicit in the model. Such an expansion of sectors of final demand is being made in a study following the general outlines of the model developed here, being conducted currently by Professor Charles M. Tiebout.

statistical discrepancy). This, too, is solely a strategic decision. A meaningful estimate of inventory change would require an inventory valuation adjustment, and since most regional accounts studies are structural rather than of a time series nature (the unbiasing of which is the major function of the valuation adjustment), it does not seem that the estimation of the adjustment should receive a very high priority.

The question of what is and what is not investment as opposed to consumption (i.e., consumer durable goods) will be considered as beyond the scope of this inquiry. Operationally, private investment consists of purchases of items by business which are expensed over more than one year plus whatever a business or individual needs a building permit to do.

Corporate profits before taxes will be credited entirely to their head office city, with the head office making all corporation income tax and dividend disbursements. With respect to the latter items, this convention makes no difference as the taxes will wind up in Washington and the dividends at the residences of the stockholders, anyway. It does leave all undistributed corporate profits at the head office, though. This is, admittedly, a bit queer, but it is at least sufficiently glaring to put the analyst on guard. A discussion of alternative possibilities has been made elsewhere.[14] Also, in part this issue will be taken up in Section III under the discussion of supplements to the basic accounts necessary for an analysis of area saving and investment.

Finally, direct exports will include not only shipments abroad and sales to area establishments of nonlocal government, but also sales to nonresidents coming into the area to make purchases.[15] The same is true for imports.

Only the first three differences between regional and national income and product accounts (the need for developing net or gross product by industrial origin, the suppression of the government sector, and the redefinition of exports and imports) will affect the

[14] Charles L. Leven, *Theory and Method of Income and Product Accounts, Including the Elgin-Dundee Area as a Case Study* (Ames: Industrial Science Research Institute, Iowa State University, 1958), mimeo., pp. 53-54.

[15] "Direct" exports means simply "exports" in the ordinary sense of the word. The concept of "indirect" exports will be introduced later.

fundamental account identity, which is shown for the national accounts as equation (5). While the other distinctions are definitional, they are important, and should be kept in mind throughout the following discussion. In any event, the new treatment of government and the new definition of the balance of trade can be taken care of symbolically quite easily by rewriting (5) as,

$$(10) \qquad Y_p = C + I + e - m$$

where $C$ is personal consumption expenditure, including private nonprofit institutions and current outlays of local government, $I$ is gross domestic investment, $e$ and $m$ are exports and imports as now redefined, and $Y_p$ is income produced in the area, which, because of the new definition of $e$ and $m$, must be distinguished from income received by area residents.

Taking care of determining gross product by industrial origin is not so simple in that it involves an allocation of imports to sectors of final demand, or,

$$(11) \qquad Y_p = (C - m_C) + (I - m_I) + (e - m_e)$$

where $m_C$, $m_I$, and $m_e$ represent imports ultimately destined for consumption, investment, and export production, respectively, and where,

$$(12) \qquad m_C + m_I + m_e = m\,.$$

For reasons which will become apparent shortly it will be useful to distinguish imports further, as follows:

$$(13) \qquad Y_p = [C - (m_C^1 + m_C^2)] + [I - (m_I^1 + m_I^2)] + (e - m_e)$$

where $m_C^1$ and $m_I^1$ are imports of consumption and investment goods in final form by final users, where $m_C^2 + m_I^2$ are imports of intermediate goods ultimately going into consumption and investment, and where[16]

[16] Since imports of goods for export production cannot be imported directly by final users there is no need to break down $m_e$ further.

$$(14) \qquad m_C^1 + m_C^2 = m_C$$

and

$$(15) \qquad m_I^1 + m_I^2 = m_I \,.$$

Quite frankly, equation (13) is a mess. Any attempt to implement it empirically would involve not only a complete set of import statistics by commodity, but also a complete interindustry analysis to properly assign the intermediate imports to sectors of final demand. While this might be a very worthwhile exercise on many scores, it represents an enormous amount of work and is not necessarily required for an analysis of area income and product. To be sure, if something equivalent to allocating imports to sectors of final demand is to be accomplished, some kind of interindustry analysis will have to be anticipated, but perhaps it can be of a much simpler variety, at least from an empirical standpoint. A start in this direction can be made by looking at a relationship which we have not yet exploited, namely, the alternative definition of income.

$$(16) \quad Y_R = Y_P - F_{xz} + F_{zx} - T_{xz} + T_{zx} - U_{xz} + U_{zx} - W_x + B_{xx}$$

where, $Y_R$ is income received by residents of the region,[17] $F_{xz}$ is

[17] $Y_R$ would differ further from the national income concept of disposable personal income in that it would not include: (1) that portion of interest earned by financial intermediaries on investments *within* the region which is paid out to *resident* investors, as opposed to being used to cover the intermediaries operating costs and, (2) taxes to and transfers from local government.

If the United States were divided into $N$ mutually exclusive and exhaustive regions the reconciliations between the summations of gross area product and area disposable personal income over these $N$ regions and gross national product and U.S. disposable personal income would be

$$\sum_{x=1}^{N} Y_P^x = GNP - G_{T_a}$$

$$\sum_{x=1}^{N} Y_R^x = DPI + \sum_{x=1}^{N} (g_{T_a}^x + g_{T_b}^x - g_{T_{tx}}^x) - \sum_{x=1}^{N} (\not{I}_{xx}^x + \not{I}_{zx}^x + \not{I}_{xz}^x)$$

where $G_{T_a}$ is total receipts of indirect taxes by federal and state governments, $g_{T_a}$ and $g_{T_b}$ are payments of indirect and direct taxes to and $g_{T_{tx}}$ receipts, by area residents only, of transfer payments from local government (transfer payments by local government to outsiders, i.e., interest on local government debt would be in $F_{xz}$), and $\not{I}_{xx}$, $\not{I}_{zx}$, and $\not{I}_{xz}$ represent interest paid by domestic financial intermediaries locally and interest received from and paid to abroad by domestic financial intermediaries, respectively.

factor payments from the region, $x$, to $z$, the Rest of the World, $F_{zx}$ is factor payments received from the Rest of the World, $T_{xz}$ and $T_{zx}$ are taxes and transfers paid to and received from the Rest of the World, $U_{xz}$ and $U_{zx}$ are unilateral payments and receipts to and from abroad, $W_x$ is undistributed profits of corporations headquartered in the region and $B_{xx}$ is business transfer payments from area enterprises to area residents. By way of explanation, undistributed profits accruing to, but not earned in, the area would have come in as part of $F_{zx}$, while area business transfer payments to nonresidents and nonarea business transfer payments to residents would be included in $U_{xz}$ and $U_{zx}$.

While (16) does not look much simpler than (13), as an empirical proposition it has a good deal more appeal. It substitutes a rather simple business questionnaire and interviews of a sample of households for a full interindustry analysis and a truck, train, boat and airplane intercepting cordon around the city. Some further analysis is called for, however, before we can say that the import assignment problem has been solved. But if we estimate $C$, $I$ and $e$ from the business questionnaire, we can write down and implement the regional income and product accounts in their simplest (one account) form, namely,

$$(17) \qquad V_P = C + I + e - (m_C^1 + m_C^2 + m_I^1 + m_I^2 + m_e)$$

where $V_P$, value added in production in the area, is equal to $Y_P$, income produced in the area, under our modified definition of "value added," and where only the sum of all of the items in the parentheses can be derived as a residual. This is not a very impressive set of accounts, but it is a start.

A really useful set of accounts can be derived if some way can be found to partition $V_P$ into $V_C$, $V_I$, and $V_e$—the value added in the production of goods for consumption, investment and export, respectively—where

$$(18) \qquad V_P = V_C + V_I + V_e\,.$$

However, before discussing this partitioning (which is the subject of the next section) it might be well first to write down the accounts which could result from it.

*Consolidated Income and Product account*

$$(19) \qquad V_C + V_I + V_e = C + I + e - (m_C^1 + m_C^2) - (m_I^1 + m^2) - m_e$$

*Consumption account*

$$(20) \qquad V_C = C - (m_C^1 + m_C^2)$$

*Investment account*

$$(21) \qquad V_I = I - (m_I^1 + m_I^2)$$

*Rest of the World account*

$$(22) \qquad V_e + (F_{zx} - F_{xz}) + (T_{zx} - T_{xz}) + (U_{zx} - U_{xz}) - (m_C^1 + m_C^2) - (m_I^1 + m_I^2) = r_{zx}$$

where $r_{zx}$ is net investment in the Rest of the World by Region $x$.

*Saving and Investment account*

$$(23) \qquad I = (W_x - B_{xx}) + (Y_R - C) + G_I - r_{zx} + d_x$$

where $G_I$ is investment outlays in Region $x$ by nonlocal government and $d_x$ is capital consumption allowances and statistical discrepancy.

The items in the first three accounts should be self-explanatory. In the Rest of the World account, factor payments, tax and transfer payments, and unilateral transfer all are shown on the left-hand side as net receipts from abroad. In the Savings and Investment account $(Y_R - C)$ is personal saving. It would include saving of households and nonprofit institutions, as ordinarily defined, and also, the current deficit (+) or surplus (−) of local governments.

Since they involve no charge against area saving, capital outlays of area establishments of nonlocal government must be included on the right-hand side of the Saving and Investment account. Alternatively, gross domestic investment could have been defined as $(I - G_I)$, but since $G_I$ does represent the physical placement of capital within the boundaries of the area it seemed preferable to keep $I$ on a gross basis. This is consistent with the treatment of employees of nonlocal government establishments in the area.[18] Business saving would be $[(W_x - B_{xx}) + d_x]$. This would represent saving only by private, for-profit business enterprises headquartered in the area, including undistributed profits earned by outside branches of locally headquartered firms. Undistributed profits earned by area branches of nonresident firms would flow out to the headquarters city as part of $F_{xz}$, and any surplus or deficit of local government enterprises would be considered separately from the business activities of the enterprise. Such deficits or surpluses would go into personal saving, being treated as interpersonal transfers between private households and a "dummy" local government agency. Some of these peculiarities in the Saving and Investment account will be covered in the discussion of the analysis of regional saving and investment in Section III.

A large part of the empirical requirements of the accounts framework outlined above can be fulfilled by means of rather simple, though fairly extensive, business questionnaires and household interviews. Items which can be obtained more or less directly from the surveys are: $e$, $m_C^1$, $m_I^1$, $F_{zx}$, $F_{xz}$, $T_{zx}$, $T_{xz}$, $U_{zx}$, $U_{xz}$, $W_x$, and $B_{xx}$. In addition, $C$ and $I$ can be determined by,

$$C = C_{xx} + m_C^1 \tag{24}$$

and

$$I = I_a + I_{bxx} + m_1^1 \tag{25}$$

where $C_{xx}$ is personal consumption expenditures on goods purchased domestically, $I_{bxx}$ is expenditure on business equipment purchased domestically, and $I_a$ is all construction. The first two of

[18] See discussion on p. 156.

these items come directly out of the surveys, while $I_a$ can be derived from secondary data.[19]

Estimates for $V_C$, $V_I$, and $V_e$ can be obtained from an analysis of $V_P$ which will be discussed in the following section. An estimate of $V_P$ could be made either from a direct survey of area establishments, or could be derived from employment data, information on "value added by manufacture" in the Census of Manufactures, and figures for "national income by industrial origin" from the National Income Statistics, depending on the amount of research resources and the extent of access to Census data. In the accounts appearing in the Appendices a combination of both methods was used.

Finally, $G_I$ poses no problem since it is equal to the nonlocal government component of $I$, by definition, and $Y_R$ can be derived from equation (16) which requires no information not already considered. This then completes the system, reducing the accounts to a set of five equations in five unknowns: $m_C^2$, $m_I^2$, $m_e$, $r_{zx}$, and $d_x$. Actual determination of the three import items would permit, of course, the estimation of the statistical discrepancy for each of the first three accounts, individually. A value for capital consumption allowances would give an independent estimate of the statistical discrepancy in the Saving and Investment account, whereas a similar estimate for the Rest of the World account would entail an elaborate analysis of foreign and domestic ownership of domestic assets and domestic ownership of foreign assets. An analysis of any one of these would probably involve more effort than that required for the whole study as presently conceived, but they can be avoided by making use of the double-entry properties of the accounting system. With these five items as residuals, we can still complete the accounts and undertake a full-scale analysis of income and output, using traditional national income analysis tech-

[19] The lack of symmetry in estimating methods for $C$ and $I$, together with the difficulty of identifying the domestically sold output of a particular establishment (a lumber mill, for example), as belonging in one or the other sector of domestic final demand could make for an inconsistency between the ratios of $V_C/V_I$ and $C/I$. This could lead to a bias in the estimate of $m_C^2/m_I^2$ but it would not affect the values of $(V_C + V_I)$, $(C + I)$, $(m_C^2 + m_I^2)$, or, more importantly, $e$, $V_e$, or $m_e$.

niques.[20] Moreover, in at least one respect the regional accounts will permit a more rigorous analysis than the national accounts, namely, that the multiplier effect of a change in final demand will take into account differences in import input requirements depending on the composition of the change.

## *II. INTERINDUSTRY RELATIONSHIPS IN REGIONAL INCOME AND PRODUCT ACCOUNTS*

The successful development and implementation of regional accounts depends on the partitioning of $V_P$ into $V_C$, $V_I$, and $V_e$. How this can be accomplished is the subject of this section. For convenience, we will consider the sectoral allocation of the value added in production for the *i*th enterprise, $v_p^i$, where there are *n* enterprises in the area. It will be assumed that any share of the firm's sales will be associated with the same share of its value added, or, in other words, that individual enterprises are single product producers with linear homogeneous production functions. This assumption is standard enough in interindustry analysis so that it need not be discussed here.

There is no problem in designating the sectoral assignment to that part of a firm's output sold to final users. The definitions in the preceding section unambiguously enable the distinction of export from domestic consumption and investment. Accordingly, a preliminary breakdown of $v_p^i$ can be written as,

$$(26) \qquad v_p^i = v_C^i + v_I^i + v_e^i + \sum_{j=1}^{n} {}^i v^j \quad (i = 1, \ldots\ldots, n)$$

[20] That a study for a single period would yield only "average propensities" might be raised as a problem in the application of multiplier analysis, since such an application implies a knowledge of "marginal propensities." However, where the increment to gross area product is accompanied by an increase in area population, as opposed to an increase in returns to factors of production, the *area's* "marginal propensity" (to consume, for example) is likely to be precisely the "average propensity" of the individuals in the area, especially if the increment to population can be considered to resemble a cross-section of the original population. See Charles M. Tiebout, "Community Income Multipliers: Population Growth Model," *Journal of the Regional Science Association,* Vol. 2, No. 1, p. 78.

where $v^i_C$, $v^i_I$, and $v^i_e$ are value added in production of goods sold directly to final users for consumption, investment, and export, respectively, and $\Sigma {}^iv^j$ is the value added in the production of intermediate goods by firm $i$ for sale to firm $j$.[21]

The last term in (26) is an $(n \times n)$ matrix showing *certain* flows from each industry (firm, in this particular instance) to every other industry. Although somewhat different from more conventional interindustry accounts, it is an interindustry account, nevertheless. Having to deal with such a set of interindustry relationships necessarily is involved in implementing regional income and product accounts, especially where they are derived from hypotheses concerning the functioning of regional and urban economies and from the public policy environment peculiar to the study area, as opposed to an accounts framework borrowed from another environmental setting, for example, the national economy. At the same time it points out that the question of consolidation of income and product and interindustry accounts at the regional level is not merely methodologically convenient, but analytically essential. Fortunately, in many cases there is reason to suppose that simpler techniques than are ordinarily used may be adequate for handling regional interindustry relationships. Before turning to these possibilities, however, the distinctions between the last term in equation (26) and more common forms of interindustry accounts will be considered.

In the interindustry accounts for the United States, in the column for each industry there is shown the number of dollars' worth of inputs from every industry, the number of dollars' worth of factor and tax payments, and the number of dollars' worth of imports going into the output produced by the given industry.[22] Looking at the table alone, one cannot tell whether the true technological relationships are obscured by the varying importance and composition of imported inputs from industry to industry. However, in making up the table,

[21] Since the analysis will be applied, ordinarily, to industry groups rather than single enterprises, ${}^iv^j \geqq 0$ for $i = j$.

[22] In particular see D. Evans and M. Hoffenberg, "Interindustry Relations Study for 1947," *Review of Economics and Statistics*, 34, No. 2, Table 4, p. 143.

> Competitive imports are charged to consuming industries through the domestic industry where the counterpart domestic production takes place. They are, thus, added to the counterpart industry's output and the imports of that industry are increased by a corresponding amount, representing the allocation from the foreign trade sector [imports]. Non-competitive imports are charged directly to the industry where they first enter domestic production. . . .[23]

Thus, the interindustry part of the table (the fifty rows and fifty columns in the upper left-hand section) represents the actual technological relationships and, accordingly, is referred to as an input-output table. Moreover, while such information is not derivable from the table itself, the proportion of each of the input requirements of each industry which is obtained from domestic sources could easily be derived from the calculations used in deriving the table, as indicated by the above quotation. These figures are not shown explicitly merely because of the relative unimportance of foreign trade for the U.S. economy. Thus, the accounts contain information on both the technological requirements for intermediate production (input-output relationships) *and* the ratio of domestic to foreign supply of each of them (trade relationships).

In most regional interindustry studies only one of these sets of relationships has been incorporated into the accounts in an essential way, namely, the input-output relationships.[24] The common thread running through these studies is that they start with a regional technological coefficients matrix, either by assuming that the coefficients are the same as for the United States, or by specifically adjusting the national coefficients. Total input (and output) require-

[23] *Ibid.*, p. 97.

[24] See for example Walter Isard and Robert Kuenne, "The Impact of Steel upon the Greater New York–Philadelphia Urban-Industrial Region," *Review of Economics and Statistics*, November, 1953; Frederick Moore and James Petersen, "Regional Analysis: An Interindustry Model of Utah," *Review of Economics and Statistics*, November, 1955; and Werner Z. Hirsch, "An Application of Area Input-Output Analysis," *Papers and Proceedings of the Regional Science Association*, Vol. V. One recent study which does incorporate trade as well as technological relationships is Roland Artle, *Studies in the Structure of the Stockholm Economy* (Stockholm: Business Research Institute at the Stockholm School of Economics, 1959). Its cost, however, was well beyond that of contemporary metropolitan studies in the United States.

ments for each industry then are computed from the input-output relationships alone. The per cent of each input that will be supplied domestically is a separate step and involves a side calculation which rests on an analysis independent of the interindustry accounts proper.[25] In essence the input-output relationships have been abstracted from the trade relationships.

Getting back to the interindustry relationships arising in regional income and product accounts, that is, the last term in equation (26), the matrix $\Sigma^i v^j$, has precisely the opposite emphasis from that of the studies described in the preceding paragraph. In recording the actual amount of each domestic firm's (industry's) output (actually, it records the value added in producing the output, but this distinction is not essential here) going to every other domestic firm, it abstracts trade relationships from input-output relationships. For want of a better name we might call it a "from-to" table.

To be sure, both from-to and input-output techniques are incomplete. Where both can be incorporated into the analysis (i.e., where regional interindustry accounts can be constructed on the same basis as those for the United States in 1947) this clearly would be desirable, and there would seem to be no basis for methodological dispute. For all practical purposes, however, this is not a realistic possibility. Thus, it seems in order to discuss their respective merits.

At the outset, it should be emphasized that we cannot say that one approach is "right" and the other "wrong." Which is appropriate depends on the questions directed to it and the structure of the region for which they are constructed. In deriving inputs of intermediate goods required by each industry under a projected change in the final bill of goods, input-output focuses primarily on the stability of the technological coefficients, while from-to

[25] Probably the most commonly used technique is that of "localization coefficients." For a description of this concept see, P. Sargeant Florence, *Investment Location and Size of Plant* (Cambridge: Cambridge University Press, 1948). For a criticism of their use in this context see Leven, *op. cit.*, pp. 12-15.

focuses primarily on the stability of the trade coefficients.[26] On deciding which approach is most appropriate for a given region, the question which really must be answered is whether the pattern of domestic supply of intermediate commodities is dependent primarily on changes in technical production functions or on the spatial distribution of production. Of course, it is always dependent on both and, to a considerable extent, the set of production functions actually employed and the spatial distribution of production are functionally interdependent, which is to argue for the superiority of a more inclusive interindustry account. But in the context of having to choose from two necessarily limited ones, some judgment as to this question must be made. While sufficient evidence is hardly available, and not likely to be forthcoming soon, and while particular cases might easily fall in one category or the other, it is hard to avoid at least a tentative judgment that from-to analysis frequently would be more appropriate, even for the largest metropolitan areas and for regions of considerable spatial extent.

Clearly, there is some limit to where from-to might be preferable (for the United States as a whole input-output, even the partial variety, would clearly be more appropriate), and this would probably be somewhere near the point where a region became big enough (in area or population) so that trade with the rest of the world accounted for less than half of its production and/or final domestic demand. For regions with a greater dependence on "foreign" trade there would seem to be little point in using input-output, since most of the intermediate goods would be supplied from abroad; for regions with less dependence from-to analysis would be of limited value, since most of the effects on suppliers of intermediate goods would be contained within the area.

This discussion of from-to vs. input-output to some extent is in contrast to views previously expressed by the author.[27] Earlier it was supposed that income and product accounts involving from-to

[26] This is meant to indicate a difference only in the focus of attention. Using either technique, the "other" consideration could be taken into account, but only by a subsidiary analysis outside the accounts model itself.

[27] For example, see Charles L. Leven, "A Theory of Regional Social Accounting," *Papers and Proceedings of the Regional Science Association*, Vol. 4, pp. 236-37.

analysis were appropriate only where a hierarchical assumption with respect to the from-to matrix (that it is triangular) was valid with minor exceptions. In practice this would seem to have limited such models to metropolitan communities of probably not more than one million population and areas of not more than a dozen or so counties, at the outside. However, on closer examination it can be seen that the hierarchical assumption is not critical with respect to the appropriateness of from-to as opposed to input-output analysis, but only with respect to the use of a simple technique (which will be discussed below) for inverting the from-to matrix. Clearly, from-to matrices can be inverted by ordinary techniques just as easily as can input-output matrices. Therefore, the author now regrets having taken his earlier position. It seems clear that the range of application of the regional income and product accounts model herein developed is much greater than was previously supposed.

As noted above, the derivation of a "direct and indirect requirements" matrix from the basic from-to matrix can be accomplished quite simply under the assumption that the original from-to matrix is triangular. Substantively, this amounts to an assumption that intermediate goods do not flow from any given industry to an industry which, in turn, supplies it; in other words, that the pattern of from-to relationships is hierarchical. In terms of equation (26) this would mean that ${}^{i}v^{j} = 0$ for $i > j$. Also, it should be noted that it is not necessary that this condition hold rigorously, but the exceptions would have to be few in number and of relatively small magnitude.

There is good reason to suppose that the hierarchical assumption is likely to be valid for most regions with populations up to at least a few hundred thousand and which are not too much bigger than a SMA.[28] To be sure, some interindustry flows commonly will be found in such regions, even where they are highly specialized. But the typical suppliers of intermediate goods in such cases would be unlikely to be supplied by their customers; the intermediate goods supplied would be mainly things like containers, business, repair

[28] See Leven, *Theory and Method of Income and Product Accounts*, previously cited.

and financial services, ice, ice cream, carbonated beverages and other food products, wholesaling services, fuel, and transportation services. Some feedbacks would certainly occur, but it is difficult to think of common examples which would not be rather trivial, like nominal purchases of manufacturers or wholesalers of minor supply items from retailers who sell their products, or telephone service provided to the light company which provides power for illuminating the telephone company offices. Some more dramatic possibilities come to mind in construction (like a contractor building a new wooden building for a lumber yard) and producer's durables, but these would be infrequent and, more important, since investment is treated as an exogenous sector any such feedback would be excluded from re-entry into the model anyway. Finally, an interrelated complex within manufacturing could occur in almost any region and would pose a real problem. However, these, too, are not likely to be found very often and where they represent a single isolated case the industries involved could be aggregated, although this could hardly be done uncritically.

Some very limited empirical findings which would *tend* to substantiate the foregoing speculations with respect to the pattern of from-to relationships are the ratios of value added in the production of intermediate goods for use by export producers ("indirect" exports) to value added in the production of goods for "direct" export in the Elgin-Dundee (Illinois) and Sioux City areas of 0.04 and 0.11, respectively.[29] Also, in conducting field work in connection with metropolitan economic studies, the Federal Reserve Bank of Chicago found little evidence of from-to relationships in manufacturing industries.[30]

As pointed out earlier, the validity of the hierarchical assump-

[29] *Ibid.*, p. 267, and unpublished analyses of the Sioux City Planning and Zoning Commission underlying Sioux City Plan and Zoning Commission, *Sioux City Economic Report,* 1959.

[30] Unpublished analyses of the Federal Reserve Bank of Chicago underlying "Growth and Prosperity in Five Midwestern Cities" in Federal Reserve Bank of Chicago, *1956 Annual Report of the Federal Reserve Bank of Chicago,* 1956.

tion with respect to the from-to matrix is entirely separate from the issue of from-to vs. input-output. In the two studies cited in footnote 29 this assumption seemed legitimate and was employed, but this was a matter of operational convenience only. If a region's input-output matrix could be assumed triangular, it would provide a similar convenience with respect to accounts developed out of input-output relationships. Since technological relationships are not substantially different for individual establishments in industry groups than for the industry itself, it might appear that this assumption for input-output matrices would probably not be valid in many cases. However, this is not necessarily the case, since the size of the input-output table relevant to a region would be smaller than that for the United States, provided that noncompetitive imports were handled outside the input-output matrix, as suggested by Evans and Hoffenberg.[31]

The usefulness of the hierarchical assumption is that, for an intermediate product, we can "look" down the chain of industry-to-industry (as opposed to interindustry) relationships and "see" its destination in some sector of final demand. For example, suppose that a region contains an ice cream plant which exports 50 per cent of its output, sells 30 per cent to local grocery stores and 20 per cent to local restaurants. The value added in producing the 50 per cent of its output sold abroad can be assigned directly to final demand for exports of ice cream. Now, suppose that 10 per cent of grocery sales are to the Rest of the World; 3 per cent (10% of 30%) of the ice cream plant's value added would be assigned to the *indirect* export of groceries and 27 per cent to the indirect production of groceries for domestic consumption. Finally, if 15 per cent of restaurant receipts come from nonresidents, 3 per cent (15% of 20%) of the ice cream plant's value added would be assigned to the indirect production of restaurant meals for export, and 17 per cent to the indirect production of restaurant meals for domestic consumption. Summarizing, the total value added *in the ice cream plant only* would be distributed as follows:

[31] *Op. cit.*

| | |
|---|---|
| Indirect domestic consumption of retail groceries .... | 27% |
| Indirect domestic consumption of restaurant meals .... | 17% |
| Indirect export of retail groceries .............................. | 3% |
| Indirect export of restaurant meals ............................ | 3% |
| Direct export of ice cream ........................................... | 50% |
| Total Value Added in Ice Cream Plant ..............| 100% |

A similar procedure could be followed for every row in the matrix *independently*, to fill out the whole table. It should be noted that there will be no rows for payments to final factors, like wages, interest, depreciation and imports, since the analysis is on a value added rather than a value of output basis. This is, of course, of great strategic significance since it circumvents the necessity for the collection of import data.[32, 33] Also, it would permit an analysis of production only [omitting all of the items in equation (16) from the analysis] using employment rather than income generated as the unit of measurement, thus affording a sort of blending of Keynesian and input-output models.[34]

It is important to note that the hierarchical assumption permits the derivation of the direct and indirect requirements in a very simple, straightforward manner, and only requires information on the total value added, total sales volume, and distribution of sales (the required distribution would be: to Rest of the World, to non-local government in region, to businesses in region, by type, other) for a carefully selected sample of firms. Not only is it unnecessary to invert the from-to matrix, but its inverse can be written down directly without even deriving the from-to matrix itself.

Symbolically, this procedure could be indicated by,

[32] Except for imports going directly to final demand, $m_C^1$ and $m_I^1$, which must be estimated to complete the income and product accounts themselves.

[33] The decomposition of value added into wage and salary and all other components obviously would be of significance for analyzing the employment consequences of any output results determined by the model, but this can be done outside the model (since it is "open" with respect to households and labor services) just as well, thus avoiding an unessential complication in the description of the model itself.

[34] The author is indebted to Professor C. M. Tiebout for pointing out this feature of the model which he is making use of in the study cited in footnote 13.

(27)

$$v_p^i = \sum_{j=1}^{n} {}^i v_C^j + \sum_{j=1}^{n} {}^i v_I^j + \sum_{j=1}^{n} {}^i v_e^j + v_C^i + v_I^i + v_e^i \qquad (i = 1, \ldots, n)$$

where $\sum {}^i v_C^j$, $\sum {}^i v_I^j$, and $\sum {}^i v_e^j$ represent the value added in the $i^{\text{th}}$ firm in the production of goods supplied *directly or indirectly* for the production for final demand of the commodity produced by the $j^{\text{th}}$ firm (a single product firm under our assumptions) for consumption, investment, and export, respectively, $m$ is the number of firms and the last three terms are as defined in equation (26). In this formulation, it might be noted, the interindustry matrix is of the order $(n \times 3n)$ since, while flows can come from only $n$ firms, they can go indirectly to any of $3n$ products; each of the $n$ commodities in each of the three sectors of final demand. In practice, some of the $3n$ columns can be eliminated since some commodities could not be for consumption by definition (construction or machine tools) and many could not be for investment (shoe laces, beer, trucking, telephone service), although all could be for export.

To incorporate the simplified from-to analysis presented above into the income and product accounts, it must be aggregated into $V_C$, $V_I$, and $V_e$. This can be done in a simple, straightforward manner. For example, the value added *in the region* in the production, *directly and indirectly* of, say, the first commodity for consumption (which differs from its production for investment or export), would be the sum of the first column in the matrix ${}^i v_C^j$ and the first element in the column vector $v_C^i$. In the table in Appendix C, for example, the value added indirectly in the production of meat-packing for consumption is the sum of the second column of the interindustry matrix, namely \$37,000, the value added directly is \$187,000, the second entry in the final consumption demand vector (shown as a column vector under final demand and repeated as a row vector beneath the interindustry matrix), with total value added directly and indirectly of \$224,000. In general then, the value added in the region in the production of the $j^{\text{th}}$ commodity for consumption, $V_C^j$ will be,

(28) $$V_C^j = \sum_{i=1}^{n} {}^i v_C^j + (v_C^j) \qquad (j = 1, \ldots\ldots, n)$$

where $v_C^j$ is $(v_C^j)'$ for convenience in addition. Similarly, value added directly and indirectly in the region in the production of the $j^{\text{th}}$ commodity for investment and export, respectively, will be,

(29) $$V_I^j = \sum_{i=1}^{n} {}^i v_I^j + (v_I^j) \qquad (j = 1, \ldots\ldots, n)$$

and

(30) $$V_e^j = \sum_{i=1}^{n} {}^i v_e^j + (v_e^j) \qquad (j = 1, \ldots\ldots, n).$$

Finally, $V_C$, $V_I$, and $V_e$, the value added directly and indirectly in the production in the region of all commodities for consumption, investment, and export, respectively, will be,

(31) $$V_C = \sum_{j=1}^{n} \left[ \sum_{i=1}^{n} {}^i v_C^j + (v_C^j) \right]$$

(32) $$V_I = \sum_{j=1}^{n} \left[ \sum_{i=1}^{n} {}^i v_I^j + (v_I^j) \right]$$

and

(33) $$V_e = \sum_{j=1}^{n} \left[ \sum_{i=1}^{n} {}^i v_e^j + (v_e^j) \right].$$

The determination of all of the items in the regional income and product accounts shown in equations (19)–(23) has now been covered.

Of course, if a hierarchical assumption cannot be made with respect to the from-to matrix, the resolution of $V_p$ into $V_C$, $V_I$, and $V_e$ would require a more complicated procedure. First, the dollar value of imports by the $i^{\text{th}}$ industry could be determined by

(34) $$m_f^j = (s_f^j - z^j) - s_f^j \sum_{i=1}^{n} {}^i u^j \qquad (j = 1, \ldots\ldots, n)$$

where $m_f^j$ is the value of *direct and indirect* imports for the production of the $j^{\text{th}}$ industry for all sectors of final demand, $s_f^j$ is the sales of the $j^{\text{th}}$ industry to all sectors of final demand (valued net of excise taxes imposed by nonlocal government), $z^j$ is the excess of *all* tax payments, except excise taxes imposed by nonlocal government, over subsidies received by the $j^{\text{th}}$ industry and $({}^iu^j) = \frac{{}^iv^j}{({}^ib^j)^{-1}}$, where ${}^ib^j = \sum_{i=1}^{n} {}^iv^j$.

Equation (34) follows from the observation that that part of the value of final output by the region's industries which is not added in the region must be added somewhere else and, hence, imported. Next, the imports of each industry could be sectorally assigned by

$$(35) \qquad m_C^i : m_I^i : m_e^i = s_C^i : s_I^i : s_e^i$$

where the right-hand side is the same as $V_C^i$: $V_I^i$: $V_e^i$ which is simply the final demand by sector, for the $i^{\text{th}}$ industry. These latter relationships come out of the assumption of each industry's producing a single product with a linear homogeneous production function. Finally, since direct imports have already been determined,

$$(36) \qquad m_C^2 = \sum_{i=1}^{n} m_C^i - (m_C^1)$$

$$(37) \qquad m_I^2 = \sum_{i=1}^{n} m_I^i - (m_I^1)$$

and

$$(38) \qquad m_e = \sum_{i=1}^{n} m_e^i.$$

Thus with $m_C^2$, $m_I^2$ and $m_e$ known, equations (19) – (23) can be solved for the five unknowns $V_C$, $V_I$, $V_e$, $r_{zx}$, and $d_x$.

Before going on to the next section, a few comments should be made with regard to the operational feasibility of the model developed in this and the preceding section. The details of the techniques have been discussed elsewhere.[35] It is hoped that they will

[35] Leven, *Theory and Method of Income and Product Accounts*, previously cited, chaps. 4-6.

be considered reasonable. Moreover, the scattered results shown in Tables 1 and 2 at least tend to show that more or less reasonable results do appear. For larger and more autonomous regions the value of the foreign-trade multiplier does increase (the especially large value for Madison might be explained by the relatively low earnings of many export workers, state government and university employees). Also, per capita regional income, etc. as shown in Table 2 differs from per capita figures for the United States in a way that might be expected.

TABLE 1. FOREIGN TRADE MULTIPLIERS FOR SELECTED AREAS

| Area | Population (000's) | Multiplier[a] |
|---|---|---|
| Winnetka, Illinois[b] | 15 | 1.05 |
| Elgin-Dundee, Illinois[c] | 60 | 1.48 |
| Evanston, Illinois[b] | 80 | 1.10 |
| Decatur, Illinois[d] | 110 | 1.72 |
| Waterloo, Iowa[d] | 110 | 1.73 |
| Sioux City, Iowa, S. Dak., & Nebr.[e] | 110 | 1.64 |
| Madison, Wisconsin[d] | 170 | 1.88 |
| Fort Wayne, Indiana[d] | 205 | 1.74 |
| Flint, Michigan[d] | 310 | 1.36 |
| Indianapolis, Indiana[f] | 620 | 1.72 |
| Milwaukee, Wisconsin[f] | 1,180 | 1.65 |
| New York, N. Y., N. J., & Conn.[g] | 16,000 | 3.36 |

[a] For comparability computed as the reciprocal of one minus the average propensity to spend plus the average propensity to import. For use of "average" rather than "marginal" propensities see text footnote 20. If investment were treated as exogenous, values would be slightly smaller.

[b] From Charles M. Tiebout, *The Community Income Multiplier: An Empirical Study* (unpublished Ph.D. Dissertation), University of Michigan, 1957.

[c] Computed from Appendix A.

[d] Computed from unpublished analyses of the Federal Reserve Bank of Chicago underlying "Growth and Prosperity in Five Midwest Cities" in *1955 Annual Report of the Federal Reserve Bank of Chicago*, 1956.

[e] Computed from Appendix B.

[f] Computed from "Big Thriving Economies" in *1956 Annual Report of the Federal Reserve Bank of Chicago*, 1957. See Tables on pp. 20 and 40.

[g] Very rough calculation based on estimate in *Made in New York* (Cambridge: Harvard University Press, 1959) that two-thirds of manufacturing is for export *and the following assumptions:* (1) ratio of value added per employee in manufacturing to value added per employee in nonmanufacturing is the same in New York as in Indianapolis and, (2) that ratio of export to total sales in nonmanufacturing is about 15% in New York.

TABLE 2. PER CAPITA INCOME AND PRODUCT, ELGIN-DUNDEE, SIOUX CITY, AND UNITED STATES

| Item | Elgin-Dundee 1956 | Sioux City 1958 | U. S. 1956 |
|---|---|---|---|
| Gross product | $2,230 | $1,850 | $2,460 |
| Income | 2,480 | 1,770 | 2,040 |
| Personal income | 2,050 | 1,675 | 1,950 |
| Disposable personal income | 1,800 | 1,460 | 1,710 |
| Personal consumption expenditures | 1,680 | 1,380 | 1,590 |
| Personal saving | 120 | 80 | 120 |

*Sources:* See references under Table 1.

## *III. ACCOUNTS FOR OTHER ANALYTICAL OBJECTIVES*

The regional income and product accounts derived in Sections I and II have been developed on the assumption that their major analytical purpose was to facilitate the analysis of the impact of forces outside the model on regional income, output, and employment. In this respect they might still be regarded as incomplete in that they ignore "feedback" relationships from the rest of the world as, for example, the secondary increase in exports stemming from an increase in income abroad by way of the region's increased imports. However, the concepts and definitions incorporated in the accounts as they now stand are entirely consistent with the computation of the foreign-trade multiplier, allowing for feedbacks.[36] This is so provided that such accounts were constructed for all regions, that the marginal propensities could be approximated by the average propensities (see footnote 20), and that the exports of each region could be broken down according to region of destination. In a sense this could be considered as a necessary supplement to the accounts for regions which accounted for a significant share of the national economy, namely that the usefulness of regional accounts is greater as the number of regions for which they have been formulated increases.

[36] See Lloyd Metzler, "A Multiple Region Theory of Income and Trade," *Econometrica,* Vol. 18, No. 4. The model developed here also seems to be consistent with an interregional model formulated in Richard Stone, "Social Accounts at the Regional Level: A Survey," Cambridge, England, April 1960 (mimeo).

However, analytical objectives other than with respect to income, output, and employment might be relevant to important policy considerations at the regional level. Some comments on the kinds of modifications or supplements to the accounts necessary to accommodate three such objectives, namely the analyses of land-use requirements in a metropolitan area, the impact of local government, and regional capital self-sufficiency, will be included in this section.

As with other primary factors of production, there is no explicit provision for land services in the interindustry section of the model; implicitly they are included in the "direct inputs" rows which do not actually appear since the table is on a "value added" basis. This presents a problem in determining the change in land requirements associated with any particular change in final demand. If one could assume that production functions were linear and homogeneous with respect to land inputs there would be no problem.[37] The output results for each industry could simply be multiplied by the appropriate land-use coefficient for each industry. However, the assumption of linear homogeneity is particularly untenable with respect to land.

As an alternative to this assumption it would be necessary to turn to a nonlinear interindustry analysis, or at least to make some kinds of *ad hoc* adjustments to the from-to coefficients. But either of these approaches would require that "land" be put into the interindustry matrix. Arithmetically, this could be accomplished by providing for a "land" row, entering in this row the rental value of the land used in the production of each of the commodities in the respective columns, and converting all of the other entries to value added exclusive of rent. But land requirements as measured by dollars' worth provide very little information about the quantity of land needed; in fact, at least in the context of preparing a master plan, it is the quantity of land to be utilized and its location which will largely determine rental values. This difficulty might be solved by substituting ${}^{i}L^{j}$, the amount of land required by the

[37] In this particular case we can use the term "production function" with respect to a from-to table, since land services cannot possibly be imported for use in the region.

$i^{\text{th}}$ industry, for the production of goods sold to the $j^{\text{th}}$ industry, in place of $({}^{i}v^{j}_{C} + {}^{i}v^{j}_{I} + {}^{i}v^{j}_{e})$. All three of these items can be replaced by a single one [thus reducing the interindustry part of the from-to table from a $(3n \times n)$ to an $(n \times n)$ matrix], since the amount of any intermediate product needed for any other product is independent of the sectoral destination of the product of the receiving industry, and since there will be no need (assuming the income and product model is already developed) to consolidate the ${}^{i}L^{j}$ into $L^{j}_{C}$, $L^{j}_{I}$, and $L^{j}_{e}$ (defined analogously to $V^{j}_{C}$, $V^{j}_{I}$, and $V^{j}_{e}$), and then into $L_C$, $L_I$, and $L_e$ (defined analogously to $V_C$, $V_I$, and $V_e$), but only into $L^{j}$, land used directly and indirectly in the production of the $j^{\text{th}}$ commodity, and $L$, total land required for all production.

The adjustments indicated in the preceding paragraph are still deficient in at least one respect, namely that they do not include the use of land for residential purposes. This is due to the model's being "open" with respect to investment demand. Thus, without modification, land used for residential construction would all show up in the final bill of goods (in the construction row of the final investment column); but this would make it impossible for the model to generate residential land requirements; they would have to be specified independently in the final bill of goods.

A simple adjustment can take care of this, namely the "closing" of the model with respect to final demand for new residential construction. This would involve a redefinition of $I$, as investment except new residential construction, and the disaggregation of the construction row and column in the basic from-to table into "construction, other than new residential," and "new residential construction." Demand for residential repair and modernization could be left in the final demand sector since they involve no direct land requirement, private nonresidential construction can be left in final demand since the land requirements for it are already accounted for in the from-to flows (this would leave open the question of how much might be spent on new business buildings, given the size of their sites), and, finally, public construction probably should be left in final demand due to its imperfect relationship with changes in city size, but it could be combined with residential construction (for example, as a constant proportion of it) and treated the same way. With these modifications the interindustry relationships for land might be written,

$$(39) \qquad L_p^i = L_f^i + \sum_{j=1}^{n+1} {}^iL^j \qquad (i = 1, \ldots, n+1)$$

where $L_p^i$ is land used by the $i^{\text{th}}$ firm, $L_f^i$ is that part of their land required for production of goods for final use, and $\Sigma^iL^j$ is that part of their land required for production of intermediate goods. To be sure, the assumption of constant coefficients is particularly weak in this case. But in expansions of final demand coming out of or by way of increases in city size, the linear homogeneity assumption might not be so serious. Even in this case though, business use of land would hardly expand proportionally, partly because of technical nonproportionality, but also because the initial conditions would not represent optimal land utilization.[38] However, where the initial conditions could be determined it might be possible to derive fairly accurate particular (as opposed to equilibrium) solutions, even under linear assumptions.

The framework of the basic accounts, as developed in Sections I and II, abstracts from the monetary relationships between local government and area residents. In analyzing regional income and output this is probably reasonably defensible since exogenous changes in local government activity are likely to be small, with no interindustry effect, if all government activity is attributed directly to final consumption or investment demand. However, the impact of changes in the region's economy on local government itself is likely to be quite significant. But, the accounts themselves can easily be expanded to include these relationships. Equation (19) would become

$$(40) \quad V_{C_h} + V_{C_g} + V_{I_h} + V_{I_g} = C_h + C_g + I_h + I_g + e - (m^1_{C_h} + m^2_{C_h}) \\ - (m^1_{C_g} + m^2_{C_g}) - (m^1_{I_h} + m^2_{I_h}) - (m^1_{I_g} + m^2_{I_g}) - m_e$$

[38] An identical model also could be constructed for amount of floor space, rather than land required.

where $C_h$ and $I_h$ are consumption and investment, other than local government, and $C_g$ and $I_g$ are the current and capital outlays of local government, respectively.

Equations (20) through (23) could be rewritten accordingly, and two additional equations inserted for the two governmental sectors. However, the efficacy of the present model for handling changes in local government activity depends on the extent to which local government output can be regarded as intermediate business product.[39] In this regard, the pertinent consideration is whether the level of any particular local government activity would, to an important extent, vary with the level of any of the other activities in the from-to matrix. In this context, who pays the taxes to finance these government functions clearly is beside the point. Education, for example, will vary importantly with population and its age distribution, but, given population, it is unlikely to be affected very much by changes in the level of a particular business activity. In short, any attempt to take it out of final consumption demand would seem a little unrealistic. The same can be said of most other services with a few possible exceptions like fire protection (chemical industry) or traffic management (increase in truck transportation). Capital spending probably should be kept in final demand also. In any event, if situations can be found where provision of intermediate business product is regarded as an important function of local government, the from-to matrix can easily be expanded to take account of this by rewriting it as

$$(41) \qquad v_p^i = v_C^i + v_I^i + v_e^i + \sum_{j=1}^{n+q} {}^i v^j \qquad (i = 1, \ldots\ldots, n+q)$$

which is the same as equation (26) but provides for an additional $q$ productive activities, the $q$ functions of local government which can be regarded as partly varying with the output of particular industries. Also, those government services which vary directly with household formation could be distributed among industries according to employment.

[39] That is, besides local government enterprises which were already included in the from-to matrix in the basic accounts.

Finally, a third policy area, with respect to which modifications in the basic accounts could be considered, is that dealing with regional capital self-sufficiency.[40] The extent to which a region is able to finance its own investment internally as opposed to seeking recourse to the national capital market certainly has significant implications for regional growth and stability, if not for maintenance of adequate regional final demand.[41] In general, to accommodate such an analysis for a region some kind of frontal attack must be made on gross flows on capital account, involving some kind of sources and uses of funds analysis. In its barest form a model for two sectors only could be represented by

$$I_{xx} + I_{zx} = S_{xx} + S_{zx} - r_{zx} \tag{42}$$

with

$$I_{xx} = S_{xx} \tag{43}$$

where $I_{xx}$ and $I_{zx}$ are locally and nonlocally financed investment (except by nonlocal government) and $S_{xx}$ and $S_{zx}$ are local and nonlocal placement of saving, with

$$I_{xx} + I_{zx} = I - I_G \tag{44}$$

and

$$S_{xx} + S_{zx} = (Y_R - C) + W_x + B_{xx} + d_x. \tag{45}$$

To make any sense, of course, each of these flows would have to be broken down in some detail for particular kinds of institutional groups, perhaps much along the lines of the Flow of Funds classification of transactors.[42]

[40] For additional remarks on this question see Leven, *Theory and Method of Income and Product Accounts*, previously cited, pp. 63-67, and "A Money-flows Analysis of Metropolitan Saving and Investment," *Papers and Proceedings of the Regional Science Association*, Vol. 8.

[41] Whether it is better to be capital "self-sufficient" or "dependent" is, of course, open to question. Nevertheless the qualitative and quantitative aspects of the relationships certainly are relevant.

[42] See Board of Governors of the Federal Reserve System, *Flow of Funds in the United States, 1939-1953*, Washington, 1955.

## *IV. REGIONAL ACCOUNTS AND NATIONAL ACCOUNTS AND RELATED DATA*

There is no question that the empirical problems with respect to constructing regional accounts of the kind discussed in this paper would be materially simplified by regional disaggregation of national accounts and other national data. Even free access to information already collected would be of considerable help. For example, if information on value added by manufacture in the Census of Manufactures could be obtained for states and metropolitan areas, by firm or at least by 4-digit standard industry classification, accurate figures on value added by activity could be derived without extensive field work or estimating procedures of questionable accuracy. Similar benefits in the case of nonmanufacturing activities would stem from free access to the Census of Business, or even from estimates of national income by industrial origin by a finer industrial breakdown. But while these developments would permit estimates of regional accounts items of a given reliability with somewhat less involved interviewing techniques, they would not materially reduce the magnitude of field work required.

It would seem that about the same observations could be made with respect to regional disaggregation of the major items in the national income accounts themselves.[43] To be sure it would be extremely useful to have regional estimates of gross national product, net national product, national and personal income, and capital, interest, and gains, as control totals and to eliminate some items on the questionnaires, but again, these disaggregations would not seem to solve the basic empirical problems of regional accounts. This would require a fundamental change in the scope of federal data.

[43] Regional double-entry accounts could not be obtained from the national accounts data alone and, moreover, the kind of supplementary information that would be needed to do this would be substantially that required for implementing the basic regional income and product accounts developed in Sections I and II of this paper. In such an event, whether we regard the regional accounts as disaggregations of the national accounts, or as constructed "from the bottom up" and made to reconcile with the national accounts, becomes largely a question of semantics.

It is entirely conceivable that a sufficient expansion in the information collected by the Censuses of Manufactures and Business (provided there was sufficient access to the information) could reduce the data collection problem for business to minor proportions. Only a few "mop-up" operations would be required, for example, for financial institutions because of the special definitions applied to them. However, the prospects for circumventing household interviewing are somewhat more dismal. At present it seems extremely unlikely that a questionnaire of the dimensions required for households could be incorporated into the Census of Population. Moreover, studies of consumer behavior also seem unlikely to provide a solution, except in particular fortuitous cases. Studies like the Survey of Consumer Finances, for example, are far too limited in coverage, and those like the 1950 Consumers' Expenditure Study are unlikely to be concerned with the kinds of information needed for the accounts. While it certainly is to be hoped that these conclusions will prove to be in error, it would seem a little unrealistic at the present time to count on a situation much more favorable than that outlined in the last few paragraphs.

As a concluding comment, it would seem appropriate to stress again the importance of maintaining flexibility in the structure of the accounts framework and of the need for building regional accounts in the context of the region's economic and policy environment. It is also hoped that it has been made clear to the reader that this flexibility and independence can be achieved without sacrificing either interregional or regional-national reconcilability or intraregional accounts integration. As a concomitant of this kind of methodological approach, in deriving all of the estimates presented in the Appendices a bias consistently has been maintained against using estimating techniques which assume that the study region is in some sense or other like some larger region or the nation. To be sure, such an assumption has been employed at times, but only as a last resort and with considerable misgiving.[44]

It would seem, at least to the author, that using straightforward national disaggregation as the essential methodological approach

[44] For a discussion of the operational procedures used see Leven, *Theory and Method of Income and Product Accounts*, previously cited, chaps. 4-6.

to empirical implementation would seriously limit the usefulness of the accounts derived. Moreover, the avoidance of preconceptions seems even more vital with respect to the theoretical structure underlying the accounts. This most emphatically is not meant to suggest that the accounts developed here represent the only adequate formulation of regional economic processes. In fact, it is hoped that precisely the opposite conclusion will be reached by the reader. Even for an analysis of income and output, the basic accounts of Sections I and II could and should be reformulated for regions with a particular economic setting or for more specific analytical objectives. Perhaps, in some cases, a framework identical to that of the national accounts would appear most appropriate. It is earnestly hoped, however, that this will be because they have been deduced to be most appropriate to the relevant questions and not simply, like the mountains, "because they were there."

## APPENDIX A:[1] INCOME AND PRODUCT ACCOUNTS, ELGIN-DUNDEE AREA, 1956 ($000)

**Income and Product Account**

| | | | |
|---|---|---|---|
| Value added in the production of goods for export | 92,555 | Sales of goods to the Rest of the World | 220,282 |
| Value added in the production of goods for investment | 5,722 | Purchase of goods by consumers | 103,973 |
| | | Purchase of investment goods | 26,180 |
| Value added in the production of goods for consumption | 38,319 | Less: Imports of intermediate goods for production of goods for export | 127,727 |
| | | Less: Imports of final and intermediate goods for investment | 20,458 |
| | | Less: Imports of final and intermediate goods for consumption | 65,654 |
| CHARGES AGAINST GROSS AREA PRODUCT | 136,596 | GROSS AREA PRODUCT | 136,596 |

[1] These accounts differ slightly from those in Leven, *Theory and Method of Income and Product Accounts*, previously cited, pp. 216-20, due to subsequent revisions.

APPENDIX A (CONTINUED)

**Consumption Account**

| | | | |
|---|---|---|---|
| Value added in the production of goods for consumption | 72,852 | Purchase of goods by consumers | 103,973 |
| | | Less: Imports of final and intermediate goods for consumption | 65,654 |
| CHARGES AGAINST CONSUMPTION PRODUCT | 72,852 | CONSUMPTION PRODUCT | 72,852 |

**Investment Account**

| | | | |
|---|---|---|---|
| Value added in the production of goods for investment | 5,722 | Purchase of investment goods | 26,180 |
| | | Less: Imports of final and intermediate goods for investment | 20,458 |
| CHARGES AGAINST INVESTMENT PRODUCT | 5,722 | INVESTMENT PRODUCT | 5,722 |

**Rest of the World Account**

| | | | |
|---|---|---|---|
| Value added in the production of goods for export | 92,555 | Net investment in the Private sector of the Rest of the World | 47,859 |
| Excess of out-commuter wages over in-commuters wages | 24,735 | | |
| Net receipts of interests, rent, and dividends from abroad | –4,969 | | |
| Excess of gifts received from over gifts given abroad | –835 | | |
| Excess of profits of outside branches of local firms over profits of local branches of outside firms | 22,485 | | |

APPENDIX A (CONTINUED)

| | | | |
|---|---|---|---|
| Less: Imports of final and intermediate goods for consumption | 65,654 | | |
| Less: Imports of final and intermediate goods for investment | 20,458 | | |
| NET CURRENT PAYMENTS DUE TO THE AREA ON PRIVATE ACCOUNT | 47,859 | NET INVESTMENT IN THE PRIVATE SECTOR OF THE REST OF THE WORLD | 47,859 |
| Excess of direct and indirect nonlocal tax and nontax payments over transfer payments from nonlocal government | 39,160 | Net contribution to nonlocal government | 39,160 |
| NET CURRENT PAYMENTS DUE TO THE AREA | 87,019 | NET INVESTMENT IN THE REST OF THE WORLD | 87,019 |

**Saving and Investment Account**

| | | | |
|---|---|---|---|
| Purchases of investment goods | 26,180 | Undistributed profits of local corporations | 9,346 |
| | | Personal saving | 6,074 |
| | | Capital consumption allowances and statistical discrepancy | 94,878 |
| | | Capital outlays of nonlocal government | 2,901 |
| | | Less: Net investment in the Rest of the World | 87,019 |
| GROSS INVESTMENT | 26,180 | GROSS SAVING | 26,180 |

APPENDIX B:[1] INCOME AND PRODUCT ACCOUNTS, SIOUX CITY AREA, 1958 ($000)

**Income and Product Account**

| | | | |
|---|---|---|---|
| Value added in the production of goods for export | 123,413 | Sales of goods to the Rest of the World | 734,437 |
| Value added in the production of goods for investment | 6,535 | Purchase of goods by consumers | 151,646 |
| | | Purchase of investment goods | 21,690 |
| Value added in the production of goods for consumption | 72,852 | Less: Imports of intermediate goods for production of goods for export | 611,024 |
| | | Less: Imports of final and intermediate goods for investment | 15,155 |
| | | Less: Imports of final and intermediate goods for consumption | 78,794 |
| CHARGES AGAINST GROSS AREA PRODUCT | 202,800 | GROSS AREA PRODUCT | 202,800 |

**Consumption Account**

| | | | |
|---|---|---|---|
| Value added in the production of goods for consumption | 72,852 | Purchase of goods by consumers | 151,646 |
| | | Less: Imports of final and intermediate goods for consumption | 78,794 |
| CHARGES AGAINST CONSUMPTION PRODUCT | 72,852 | CONSUMPTION PRODUCT | 72,852 |

[1] These accounts differ slightly from those reported by the Sioux City Plan Commission due to subsequent revisions.

APPENDIX B (CONTINUED)

**Investment Account**

| | | | |
|---|---|---|---|
| Value added in the production of goods for investment | 6,535 | Purchase of investment goods | 21,690 |
| | | Less: Imports of final and intermediate goods for investment | 15,155 |
| CHARGES AGAINST INVESTMENT PRODUCT | 6,535 | INVESTMENT PRODUCT | 6,535 |

**Rest of the World Account**

| | | | |
|---|---|---|---|
| Value added in the production of goods for export | 123,413 | Net investment in the Private sector of the Rest of the World | 13,723 |
| Excess of out-commuter wages over in-commuters wages | (−11,160) | | |
| Net receipts of interests, rent, and dividends from abroad | 475 | | |
| Excess of gifts received from over gifts given abroad | (−428) | | |
| Excess of profits of outside branches of local firms over profits of local branches of outside firms | (−4,628) | | |
| Less: Imports of final and intermediate goods for consumption | 78,794 | | |
| Less: Imports of final and intermediate goods for investment | 15,155 | | |
| NET CURRENT PAYMENTS DUE TO THE AREA ON PRIVATE ACCOUNT | 13,723 | NET INVESTMENT IN THE PRIVATE SECTOR OF THE REST OF THE WORLD | 13,723 |

APPENDIX B (CONTINUED)

| | | | |
|---|---|---|---|
| Excess of direct and indirect nonlocal tax and nontax payments over transfer payments from nonlocal government | 37,559 | Net contribution to nonlocal government | 37,559 |
| NET CURRENT PAYMENTS DUE TO THE AREA | 51,282 | NET INVESTMENT IN THE REST OF THE WORLD | 51,282 |

**Saving and Investment Account**

| | | | |
|---|---|---|---|
| Purchases of investment goods | 21,690 | Undistributed profits of local corporations | 3,416 |
| | | Personal saving | 8,891 |
| | | Capital consumption allowances and statistical discrepancy | 59,947 |
| | | Capital outlays of nonlocal government | 718 |
| | | Less: Net investment in the Rest of the World | 51,282 |
| GROSS INVESTMENT | 21,690 | GROSS SAVING | 21,690 |

APPENDIX C: VALUE ADDED IN PRODUCTION DIRECTLY AND INDIRECTLY, SIOUX CITY AREA, 1958 ($000)

| To / From | Intermediate production for | | | | | | | | | | | |
|---|---|---|---|---|---|---|---|---|---|---|---|---|
| | Consumption of | | | | | | | | | Investment in | | |
| Industry | I | III | IV | V | VI | VII | VIII | IX | | II | V | VII |
| I Agriculture | * | * | * | * | * | 18 | * | 1 | | — | — | — |
| II Construction | — | — | — | — | — | — | — | — | | 564 | — | — |
| III Meat packing | — | — | — | — | — | 2,075 | — | — | | — | — | — |
| IV Other food manufacturing | — | — | 31 | — | — | 2,190 | — | — | | — | — | — |
| V Other manufacturing | — | 2 | 8 | 14 | 134 | 347 | 52 | 443 | | 283 | 249 | 246 |
| VI Transportation & communication | * | 23 | 44 | 44 | 458 | 1,597 | 212 | 803 | | 81 | 57 | 81 |
| VII Trade | 2 | 9 | 21 | 23 | 262 | 2,595 | 121 | 460 | | 613 | 39 | 65 |
| VIII Finance, insurance & real estate | * | 1 | 7 | 7 | 63 | 179 | 30 | 113 | | 12 | 7 | 12 |
| IX Services | * | 2 | 12 | 12 | 126 | 353 | 58 | 249 | | 24 | 17 | 24 |
| Intermediate production to | 2 | 37 | 123 | 100 | 1,043 | 9,354 | 473 | 2,069 | | 1,577 | 369 | 428 |
| X Government | | | | | | | | | | | | |
| Total Final Demand | 58 | 187 | 709 | 737 | 7,782 | 21,738 | 3,609 | 13,665 | 11,166 | 1,479 | 1,123 | 1,559 |
| Total Production of | 60 | 224 | 832 | 837 | 8,825 | 31,092 | 4,082 | 15,734 | 11,166 | 3,056 | 1,492 | 1,987 |
| | | | | | 72,852 Consumption, intermediate and direct | | | | | 6,535 Investment, intermediate and direct | | |

* Less than $500.

## Appendix C (continued)

| Ind. | Intermediate production for: Exports of I | II | III | IV | V | VI | VII | VIII | IX | | Intermediate production from | Final demand: For consumption | For investment | For exports | Total final demand | Total production by |
|---|---|---|---|---|---|---|---|---|---|---|---|---|---|---|---|---|
| I | * | * | 44 | 1 | 1 | * | 8 | * | * | | 73 | 58 | — | 221 | 279 | 352 |
| II | — | — | — | — | — | — | — | — | — | | 564 | — | 1,479 | 1,613 | 3,092 | 3,656 |
| III | — | — | — | — | — | — | 49 | — | — | | 2,124 | 187 | — | 36,088 | 36,275 | 38,399 |
| IV | — | — | — | 444 | — | — | 675 | — | — | | 3,340 | 709 | — | 10,334 | 11,043 | 14,383 |
| V | 2 | 14 | 1,389 | 49 | 748 | 62 | 593 | 38 | 184 | | 4,857 | 737 | 1,123 | 19,641 | 21,501 | 26,358 |
| VI | 8 | 90 | 4,751 | 563 | 1,068 | 261 | 1,190 | 114 | 294 | | 11,739 | 7,782 | — | 4,866 | 12,648 | 24,387 |
| VII | * | 5 | 760 | 34 | 59 | 13 | 1,097 | 6 | 15 | | 6,199 | 21,738 | 1,559 | 17,287 | 40,584 | 46,783 |
| VIII | 1 | 13 | 396 | 85 | 160 | 39 | 140 | 18 | 45 | | 1,328 | 3,609 | — | 2,082 | 5,691 | 7,019 |
| IX | 2 | 25 | 542 | 158 | 300 | 73 | 264 | 32 | 93 | | 2,366 | 13,665 | — | 5,446 | 19,111 | 21,477 |
| IP | 13 | 147 | 7,882 | 1,334 | 2,336 | 448 | 4,016 | 208 | 631 | | 32,590 | | | | | |
| X | | | | | | | | | | | | 11,166 | — | 8,820 | 19,986 | 19,986 |
| TFD | 221 | 1,613 | 36,088 | 10,334 | 19,641 | 4,866 | 17,287 | 2,082 | 5,446 | 8,820 | | 59,651 | 4,161 | 106,398 | 169,210 | |
| TP | 234 | 1,760 | 43,970 | 11,668 | 21,977 | 5,314 | 21,303 | 2,290 | 6,077 | 8,820 | | | | | | |
| | | | | | | | | | | | | | | | | 202,800 |

123,413
Exports, intermediate and direct

## COMMENT

*Charles M. Tiebout*

*University of California at Los Angeles*

Professor Leven's fine paper essentially does three policy-oriented things. It (1) sets up income and product accounts; (2) develops and applies his "from-to" matrix; and (3) considers modifications of his basic set of accounts for other analytical objectives.

The income and product accounts as developed by Leven seem satisfactory. A central problem which plagues regional as opposed to international trade studies, the allocation of corporate profits, has been solved in a manner which, as Leven puts it, is "sufficiently glaring to put the analyst on guard."

Frankly, I am willing to settle for a less elegant set of regional accounts, given the data problems. In this respect, "income accruing to residents" seems less messy. The sum for all regions would be akin to U. S. personal income. This is more in the spirit of Leven's equation (16). If the policy interest of the study is on the forces of demand affecting the region, then we would like "income accruing to residents" as the sum of income from the various sectors; investment exports and so forth. Leven's "from-to" table could be switched to read dollars of income instead of value added. Additional rows can be added to include: transfer payments into the region, dividend receipts, and other such items.

In short, to the extent that "income accruing to residents," as contracted with some "area product" aggregate, is as easy to obtain, it may be a more useful aggregate.

Leven's real contribution, and it is a contribution, comes via his from-to model, specifically Appendix C and the discussions thereof in the text.

From-to tables, I would argue, are a cross between input-output and Keynesian models—the latter often called base studies. In many ways from-to constitutes a useful blending. Why is this the case?

The data in Appendix C are in terms of dollars. Yet, we could imagine that these are the numbers of persons employed. Instead of sectoring the source of area product, one could sector out area employment. And while we all prefer dollar units for most purposes, data limitations and the scope of the study often force us to turn to employment. Thus, Leven's framework is suitable when employment units are used.

There is, of course, nothing sticky about this blend of input-output and a Keynesian model. Sectors can be added. In a California study, instead of Leven's three sectors we use seven: (1) exports to government; (2) exports to private; (3) local consumption; (4) housing investment; (5) business investment; (6) state, local and some federal current operations; and (7) state, local and some federal investment activities.

What all this amounts to can be simply stated. Leven's from-to table is a useful blend of input-output and simple multiplier analysis. The loss of some interindustry detail on the input side is more than offset, I feel, by the operational simplicity.

When it comes to modifying the basic accounts for other policy considerations, I am appreciative, but less enthusiastic.

The idea of attaching land use coefficients to an interindustry model is appealing. At the conceptual level it would seem that, in addition to land requirements, the model tells us something about transport flows. A more specific incorporation of transport flows might be introduced to indicate something about patterns of land use. For example, which industries will agglomerate near some key industry and which are more footloose within the urban scene. But Leven's discussion and this suggestion have been at the conceptual level.

Operationally, the prospect is glum. Land use forecasts needed for planning come at the tail end of a forecast. That is, having forecasted growth in exogenous sectors, the interindustry table grinds out the endogenous sectors. Given a few assumptions about population, and so forth, we now forecast land use requirements. Any error along the line compounds in the land use forecasts.

Finally, with respect to Leven's impact on government modifications of the accounts, he may be underestimating their potential contribution. To the extent that new industry, and the population it attracts, requires additional government services, there is interest in some kind of a fiscal residuum, along the lines suggested in Hirsch's paper. Not every industry and the employees it attracts, given the tax structure and "demands," adds more to the tax base than it calls forth in expenditures. A from-to table might be augmented to point this out.

These comments, as is obvious, are highly complimentary, and by intent. Leven's work offers a contribution at both the theoretical and operational levels. And, as a bonus, it really does have policy potential.

7

# REGIONAL ACCOUNT PROJECTIONS IN THE CONTEXT OF NATIONAL PROJECTIONS

*Louis Delwart and Sidney Sonenblum*[1]

*National Planning Association*

AMONG THE MOST significant advances of the last several decades in the development of tools for aiding public and private economic policy has been the improvement in the use and understanding of national economic accounts. The demonstrated utility of these national accounts has encouraged the development of subnational accounts and national and subnational account projections.

The interest in both regions and projections is not new, although pursuing these interests by expressing the relevant data in an accounting framework is relatively new. This interest is not only evident in the United States; often overlooked is the strong interest in many other countries.

[1] This paper has been written in connection with a current research program of the National Economic Projections Series of the National Planning Association. The authors are particularly grateful to NPA Chief Economist Gerhard Colm, who had the difficult task of contributing to the development of the paper and then serving as a critic. They are also indebted to Joseph L. Fisher, of Resources for the Future, and Douglas Greenwald, of McGraw-Hill, for their perceptive and constructive comments which the authors have tried to incorporate.

We shall discuss the uses and types of regional accounts, including a description of the economic indicators that are important in economic accounts. We shall then discuss the procedures for projecting these economic indicators, first summarizing these procedures and relating them to some general statements about regional development and theory, and, finally, describing the detailed procedures for making the projections.

## *I. USES OF REGIONAL ACCOUNTS*

The significance of regional accounts is that they permit the systematic organization of data in a manner conducive to the formulation of policies and the solution of problems. The pressure of current public and private policy problems indicates two main problem areas. First are the public policy problems. State and local governments have the problem of encouraging the attraction and expansion of local industry in a manner consistent with the long-run aspirations and resource availabilities of the area. Federal government activities also have an effect on differential local development. However, the extent to which the federal government should directly participate in the achievement of local aspirations in a manner consistent with national goals is subject to much discussion; an example is the debate over provision of aid to depressed areas harmfully affected by national economic growth.

On the private policy level the chief concern is with the determination of expected demands for a particular enterprise's products in specific areas, and the related problem of where new plants should be located. Put somewhat broadly, these various decisions require information about potential access to input and output markets.

Regional account projections aid in the establishment of public and private policy. Projections between five and ten years ahead are found to be the most useful to private policy according to a recent National Planning Association survey of business economists. Yet the same survey finds that projections over twenty-five years are also useful and certainly public planning for such activities as urban development, water resource development, and other capital facilities construction would often require a perspective of up to fifty years.

The state is the basic building block for our projections. It is clear that no given geographic unit is always most appropriate for economic analysis. What is being sought is a set of data which can be useful to a variety of users who have different problems and whose primary interests are in different regions. Our projected data can be conceived by users as raw materials for their further data processing and analysis. The depth analyses required, for example, by a metropolitan area study, which needs a detailed input-output table for its purposes, can be aided by the use of these projected economic indicators.

The state offers the best compromise region for these various purposes. Data are most abundant for states and potential users are most familiar with state data. (The NPA survey indicates that the region preferred by business economists for economic projections is the state.) Furthermore, states can often be subdivided so as to show different activities as being associated with metropolitan areas and their environs. Finally, states can be combined into multi-state regions, with distinct migration trends of their own, which can then be used for further analysis of the specific state activities.

## *II. REGIONAL ACCOUNTS AND SELECTED ECONOMIC INDICATORS*

Both public and private policy determination require data dealing with the national scene as well as with strictly local developments, since the local data can only be interpreted with reference to what is happening in other areas and the nation as a whole.

The necessary local data required for regional accounts will, of course, depend upon the type of local account under consideration. Since a region's activities include the same type of transactors and transactions as the nation, the four basic national accounts could have their regional counterparts.

A regional income and product account presumably shows local households, government, the total of local business, and the Rest of the World (national plus foreign) as transactors; the transactions shown are the expenditures and receipts among transactors. The regional income and product account is designed primarily to show the consolidated net income and product of the region and, if possible, the geographic origin and destination of this income and product and its relationship to the nation's income and product.

In a regional moneyflows account the transactors are local and nonlocal financial and nonfinancial units, including households and government; the transactions are the flow of funds among the various transactors. A regional moneyflows account is designed primarily to show the financial relationships between a region and the Rest of the World or the regional distribution of and changes in ownership claims of national assets.[2]

Regional wealth accounts are concerned with evaluating the reproducible assets and resources of the community. The simplest form of such an account would be a listing of the value or volume of locally situated resources, including plant, equipment, labor skills, land, raw materials, and so forth. A more difficult task is the development of wealth accounts which can be linked with current accounts, such as interindustry or income and product accounts, and to include with the asset and resource valuations appropriate valuations of nonmarket resources such as weather or access to a favorable transportation network.

In a regional interindustry account the transactions are the value of purchases and sales of goods and services among the specific local and nonlocal businesses, households, and government interests. A regional interindustry account is designed to show the trading transactions among industries and regions. A complete

[2] The national moneyflows account, although including a statement of sector claims and liabilities, does not include a complete statement of assets (omitted are corporate stocks, tangible assets and net worth). A regional balance sheet of assets and liabilities emphasizing ownership of national assets by the region's residents might therefore be an additional type of account, differing from a regional wealth account emphasizing physical assets within the region.

regional interindustry account would be both multi-region and multi-industry.

That each national account could have its regional counterpart does not mean that the regional accounts need strictly follow the national definitions. On the contrary, there are many reasons for departing from national definitions—a point which is pursued in other papers presented at the Conference on Regional Accounts.[3] It would be convenient if the regional accounts added up to the national control totals, but even this is not essential if the specific reasons for the differences are identified.

Further than this, it is not necessary that each regional account take the exact form of its national counterpart. Indeed, it is very possible that the most fruitful form of a regional account may be one which combines various aspects of the different accounts, such as the model shown by Werner Hirsch.

The projections with which we are concerned are not regional account projections. Rather they are projections of economic and demographic indicators which can be used directly in specific economic analyses or as inputs for the development of projected regional accounts which would improve economic analysis by virtue of the internal consistency offered by accounts. Given the present relatively undeveloped state of regional accounts, it is most urgent to have regionally projected values for critical economic indicators rather than to attempt the more difficult task of projecting complete accounts.

The regional indicators which we believe are most important in the analysis of current private and public policy problems are:

Population (distributed by age)
Labor force
Employment (total for region; total in each 2-digit manufacturing industry and 1-digit nonmanufacturing; in some instances still further disaggregation might be necessary)

[3] See Werner Z. Hirsch, "A General Structure for Regional Economic Analysis"; Richard Ruggles and Nancy D. Ruggles, "Regional Breakdowns of National Economic Accounts"; and Charles L. Leven, "Regional Income and Product Accounts: Construction and Applications," in this volume.

Value added (total for region and for each industry)
Employee compensation paid by each industry
Industry productivity
Disposable income (of households)
Personal income (of households)
Personal expenditures (total and broad budget groupings)
Household income distribution
Indexes of gross output for each industry
Investment expenditures
Government expenditures

The aim of our research is to project fifteen years ahead the state allocation of the list of diverse economic indicators mentioned above, including estimates for the ten 1-digit standard industry classification industries and the twenty 2-digit industries within manufacturing.

## *III. IN LIEU OF A THEORY OF REGIONAL DEVELOPMENT*

The concept of the nation's economic budget is the most useful tool for making national economic projections. The budget framework permits the implicit or explicit introduction of economic relationships determined by both macro- and micro-economic theory. There is no comparable tool for making regional projections as, indeed, there is no adequate theory of regional economic growth.

Yet, precisely because the projection of regional economic activities is still only a crude art, it is essential that we respond to the question, "What theory or theories of regional economic growth underlie this exercise in practical methodology?"[4] We have no comprehensive theory, although our methodology is largely guided by two major premises and two important simplifying assumptions.

### *The Two Major Premises*

*The Region as a Part of the Nation.* It is our judgment that regional development is *primarily* determined by growth and

[4] Asked by Joseph L. Fisher in his comments on this paper.

changes in the national environment. Of course, regions have grown and will continue to grow at different rates depending on their comparative ability to serve national and local markets—an ability which in turn determines the size of local markets. In fact, determining the manner in which regions react and what they react to is the chief difficulty in making regional projections. Yet it remains our judgment, as supported by past experience, that the range of potential reactions to given national impulses by specific regions is sufficiently limited as to permit reasonable quantitative estimates for the future.

Thus, national projections serve as the starting point for regional projections, just as national growth serves as the starting point for analyses of regional development. There are several further advantages to beginning with national projections. First, they can probably be made with more confidence than can regional projections, which introduce the additional factor of migration of people and capital. Second, the national aggregates serve as control totals which must agree with the sum of regional aggregates. This accounting constraint limits the possibility that inconsistency among different regional developments will be projected. Finally, the national totals provide a restraint which assures that similar assumptions will underlie the separate analyses of the variety of different users of projections. Our national estimates will use the judgment model projections of the national economy prepared by the National Planning Association, which are based on the broad assumptions of high employment, no drastic price inflation, and no nuclear war.

*Employment as a Point of Departure.* It is our judgment that the analysis of employment in specific industries provides the most fruitful initial approach to regional projections. This judgment is based partly on the fact that employment data are more readily available than any other regional economic data. Principally, however, it is based on our belief that the changing migration decisions of business and people—which are essentially determined by the relative advantages offered by the region in the use of their capital and skills—are adequately reflected in the data on state-industry employment trends. In this sense, past trends are crucial to an understanding of future possibilities.

By this we do not mean to imply that investment location decisions are irrelevant to regional development; rather, we believe that these decisions are implicit in our employment migration estimates. On the other hand, we do suggest that population moves largely in response to job opportunities; there are, of course, exceptions, particularly in certain age groups, and these must be taken account of. Finally, while our concern is with interstate migration, we recognize that the major part of population mobility in the past has been related to migration into metropolitan areas and suburbs. We believe that this movement will continue and is accounted for in our state industry employment analyses.

### *Two Simplifying Assumptions*

The complexity of interregional relationships has compelled us to adopt two principal simplifying assumptions which again reflect our theoretical judgments as to how regions develop without offering a theory of regional development.

*Allocation of New Capacity.* The impact of national developments on specific regions, we believe, is felt through the region's reaction to changing demand and output of key industries. Theoretically, we visualize the system as a gigantic interindustry table with forty-nine regional units (forty-eight states of the continental United States and the District of Columbia) requiring a simultaneous solution.

Statistically, however, the critical element to project is where new, rather than total, capacity will locate. Our approach is guided by the belief that there are very strong pressures for existing capacity to remain where it is, and that additional capacity, unless there are strong reasons to indicate the contrary, will tend to locate where similar capacity is already located. There is even a tendency for the capacity of new products, unknown in the past, to locate in already heavily developed regions and for footloose industries to form a location pattern of their own heavily influenced by the opportunities for amenities. It might almost be said that inputs are regionally more stable than outputs, a stability explained by such factors as the depreciation possibilities of capital, the location of raw materials, the desire of people to stay with ties already developed, and so forth.

It is possible that as transportation improves, as more rapid technological advances become associated with more rapid obsolescence, and so forth, migration will become relatively less dependent upon the national growth of key industries. However, our justification for making regional projections remains based in the belief that much of what we can observe in the current situation is similar to the situation expected to prevail some ten or fifteen years from now.[5]

[5] The magnitude of the potential judgment error in making regional projections in the context of national projections is related to the extent to which it is new rather than total capacity that is relevant to differential regional growth. This magnitude can be put into somewhat better perspective if we see how many of the estimated 21 million manufacturing employees in 1976 can be regarded as regionally fixed.

With twenty manufacturing industries and forty-nine states (including the District of Columbia) there are 980 conceivable instances in which a 2-digit manufacturing industry within a state could show an absolute decline in employment. Between 1947 and 1957 there were 350 such instances, involving one million employees, or 7 per cent of 1947 manufacturing employment.

Looking to the future, we can try to see how much of projected manufacturing employment would be reasonably distributed if we made the following extreme assumptions: first, that all state industries whose employment in the past has increased faster than the national industry average will, in the future, increase at least at the national average; second, that all state industries which have increased but at less than the national rate will, in the future, retain their initial-period level of employment; finally, that all state industries which have shown an absolute decline will, in the future, have no employment. Since these are extreme assumptions they provide a fair measure of magnitudes to which judgments are applied.

To test these assumptions we first identified the three classes of industries within states by using data for the period between 1939 and 1954. [For this purpose we used production worker data to be found in Perloff, *et al.*, *Regions, Resources, and Economic Growth* (Baltimore: Johns Hopkins Press, 1960)]. We next determined national industry employment growth rates for the period between 1947 and 1957. Finally, we projected the 1947 data to 1957 on the basis of the above assumptions.

The results indicate that, assuming "correct" projections of national industry employment growth, the indicated extreme assumptions result in a regional allocation of 80 per cent of projected manufacturing employment including 319 instances, covering 5 per cent of manufacturing employment, in which the projection for an industry within a state was higher than the actual.

Thus about 25 per cent of manufacturing employment cannot be regionally

*The Use of Economic Base Theory.* Reliance on some aspects of economic base theory is the second of our simplifying assumptions. In analyzing the demand and supply characteristics of specific industry developments in the region we find it convenient to distinguish between lead and derived industry outputs on the one hand, and final and intermediate markets on the other.

Determining which are the lead and which are the derived industries for any specific region depends on analysis of that region. In general, the lead industries are the commodity producing industries—manufacturing, mining, and agriculture. The regional lead industry projections, as already indicated, are constrained by the fact that the critical element to project is where new, rather than total, capacity will be built. The lead and derived industries can each sell to final and intermediate markets within or outside the state. Our procedure lays first and special stress on the projection of lead industries; these in turn will permit the projection of derived industry activities since the relationship between lead and derived industries is likely to vary within a sufficiently narrow range for projection purposes (even if it varies too greatly for other purposes).

We will therefore concentrate much of our effort on distributing regionally projected U.S. increments in lead industry employment. Even for the task of projecting increments, many inputs can be regarded as relatively stable. Our first job, therefore, will be to allocate employment increments on the basis of where the new

---

"fixed" by our extreme assumptions. This includes the total employment in those state industries whose employment has declined in the past (about 7 per cent of manufacturing employment); the addition to employment in those state industries which have grown in the past but at less than the national industry rate (about 5 per cent of manufacturing employment); the employment which is in excess of the national industry growth rate in those state industries which in the past have grown faster than the national industry (about 8 per cent of manufacturing employment); and, finally, that employment in state industries which our assumptions have regionally "fixed" but which are projected greater than actually result (about 5 per cent of manufacturing employment). While ample room for wrong judgments still remains, particularly in the case of individual states with rapid differential growth, it is encouraging that these judgments are circumscribed by the existence of a large proportion of permanently located employees.

plants (or the additions to old plants) will be likely to locate, if they follow what they consider to be the path of least resistance and if the set of locational pressures displays only slow change over the fifteen-year period.

## *IV. OUR APPROACH TO REGIONAL PROJECTIONS*

Before we proceed with a description of our three-stage approach, we feel it necessary to describe the manner in which we introduce into our regional projections those supply and demand factors that are deemed critical from the point of view of causing migration of industry employment and capital. Then we can take advantage of the work of regional analysts who have identified the most important supply and demand factors of location and have also uncovered a number of fruitful correlations among economic activities on a regional basis. The way that the mutual feedback among supply and demand factors is treated becomes one of the most important steps in making projections.

The mutual interaction between supply and demand factors could be viewed in a dynamic manner: the seeds of future regional development are present in current activities of the region. Current levels of investment or profit rates will have an effect on future output; current population growth will influence future employment, and so forth. The difficulty with a dynamic approach of this nature is that we do not at present know enough about the relative intensities of the various factors in operation, nor about the time interval it takes for their effects to be felt or dissipated.[6] That will not be our approach, therefore, at this time.

[6] This problem can be spelled out more completely with a simple illustration. The regional demand for labor is equivalent (*ex post* and with some statement about productivity) to the supply of regional output. Current market and profit expectations therefore could be viewed as generating a capacity increase which motivates labor to move into the affected area in amounts as predetermined by the ratio of labor to output; the labor mobility in turn generates further demand which affects expectations, and so on. However, at the same time that job opportunities induce population mobility which in turn affects industry location, nonjob-oriented mobility is affecting market size and expectations. In our observations of historical data we do not really know which piece of total mobility is job oriented and which is not; that is, we cannot even determine the sequence among labor mobility, expectations, and industry mobility, let alone the quantitative relation which defines the sequence.

Our approach is more one of comparative statics. We look at the interrelation between demand and supply factors in an accounting sense where the demand and supply factors are consistent or equal with each other (*ex post*). Following usual social accounting practices we first make projections on the basis of analysis of the supply side, and then of the demand side, and by appropriate adjustments make them finally consistent. We concentrate our major efforts on mutually consistent projections of state industry employment and income and derive the remaining economic indicators by use of suitable coefficients.

### *A Three-Stage Approach*

To simplify the handling of the vast amount of data involved in making regional projections, we have adopted an approach involving three successive stages of approximation. The first approximation model considers the trends in state industry employment and income; while introducing the element of time into the projections it also obscures the vast complexity of events and forces in each state and region. The second and third approximation models analyze the effects of specific supply and demand factors in order to adjust the first model in those instances where shifts in trends seem warranted.

*First Approximation.* Lead industry employments by state are first projected on the basis of the distinction between proportionality and differential shifts in employment. Proportionality shifts, which show the effects on state economic activities of whether the state is endowed with slow- or fast-growing national industries, are determined by our national industry growth analyses. Differential shifts, which show the extent to which a state industry is growing faster or slower than the national average for that industry, are based on extrapolations of past trends in the state industry's share of national industry employment. Such extrapolations assume implicitly that locational forces will continue to have roughly the same (as in recent years) differential impacts on the growth of state industries.

This is the essence of our so-called first approximation model. However, adjustments to the simple extrapolation of state industry employment shares will be required by analyses of trends in income, population, and total employment of states as well as analysis of the possible effects of nationwide organizational shifts such as the possible decentralizing impact of an increasingly flexible transportation system.

This first approximation model will also project derived industry employments through analyses of the trend in each state's share of the national industry employment. These are needed at this stage not so much for their own sake, but for obtaining improved projections of state population and of the relative state advantage in the supply of needed factors for industry production.

*Second Approximation.* The second approximation model adjusts the projected trends of differential growth for the lead industries by analyzing the potential effects of specific supply factors and determining how they may be expected to behave regionally in a manner different from that evidenced by past trends. Such analysis depends on determining the relative advantage different states have in supplying specific industry needs which significantly affect the industry cost structure.

The principal factors considered relate to the over-all cost position of an industry, such as the effects of transportation costs, economies of scale, threshold effects, and so forth. Also included in the second approximation is an examination of linkages among commodity industries.

Thus the first approximation projection of the state industry's supply of output is adjusted in the second approximation by an analysis of the national industry's orientation needs and the state's relative ability to supply these needs. Determination of the projected relative advantage in the supply of labor, markets, and amenities is aided by the first approximation model which allows us to obtain improved state population estimates, as well as by analyses of the data sources indicated below.[7]

The second approximation is the major step in this projection procedure. It has been suggested that the adjustments required in the second approximation may be so significant as to make the first approximation unnecessary.[8] If this were so it would suggest that the dramatic new changes that are occurring in our economy are so significant in their effects as to make the study of past trends irrelevant to judgments about the future. While such judgments should include speculation about new developments which might have marked effects on past trends they should not be based solely on such speculation.

Many of the new developments that are likely, particularly those related to the research and development field, will probably have their major effect at the national level, that is, on the distribution of national industry employment rather than on the regional distribution of a given industry's employment. However, although we can identify much of the problem of new developments as being at the national level, we cannot be confident that our national industry analyses are adequate. Improvements in national projections are critical, both for their own sake and for improving regional projections; but it should also be noted that national projections have already been proved useful.

[7] These advantages and needs and the data which permit their determination are summarized by Perloff as follows:

| *Relative State Advantage* | *Relative Industry Needs* | *Data Sources* |
|---|---|---|
| 1. Size of markets (including economies of scale) accessibility | Market orientation | Sales (including interindustry) data |
| 2. Quality and quantity of labor accessibility | Labor orientation | Manpower data |
| 3. Natural resource accessibility | Resource orientation | Stock (including marginal productivity of capital) data |
| 4. Amenity accessibility | Amenity orientation | Stock (including marginal productivity of capital) data |

[8] By Douglas Greenwald. Gerhard Colm expresses uneasiness about the same issue and provides an impressive list of factors which must be considered in the second approximation.

When we concern ourselves with how the employment effects of the national developments are regionally distributed, the task, as we have indicated, is to identify the type of locational factors that are critical for each industry and the locational advantages in each region that are already there or are likely to develop. This is certainly more speculative for new developments than for those we can already anticipate. However, even here the preliminary information on employment, population, income, and industrial structure shown in the first approximation cannot be dispensed with. We should further remember that these effects are constained by the tendency toward regional stability of inputs.

*Third Approximation.* The third approximation model will adjust the second approximation model's state industry employment estimates by consideration of regional demand factors. Thus the third approximation represents an attempt to further specify what underlies the state's relative advantage in market accessibility as related to the industry's market needs.

In the third approximation, the market factors related to commodity industries are considered in the context of the relative significance to each industry of final and intermediate markets located inside and outside the state. Thus, for example, we can introduce the effect on a given state industry's output of previously estimated markets in broad regions, or the effect of estimated state population and income on the output of a commodity industry particularly influenced by final sales within the state. Once these commodity industry adjustments are made, the industry employment estimates can then be checked against observable covariances between state population and a number of state lead-industry employments.

Also in the third approximation we obtain improved noncommodity industry employments on the basis of trends in the ratio for each state between lead and derived industry employments, supplemented by consideration of regional household income and expenditure distributions and the industry relation between employment and employee earnings.

In this same third approximation, we make a final reconciliation of the various factors by again looking at improved population estimates and their relation to commodity and noncommodity industry employments, since there are marked trends in state indexes of industrialization, decentralization, and income, all of which should be mutually consistent in the projections.

### *From Regional Employment to Other Regional Indicators*

Having derived projected matrices of state industry employment (which for forty-nine states and thirty industries require 1,470 individual and mutually compatible cells), we next derive projections of the other economic indicators by relating them to the employment estimates. Such relationships, however, will depend on analysis of additional data, such as consumer surveys, to determine personal expenditure distributions, productivity data to determine income originating and employee compensation by each industry, transfer payment data to determine disposable income, birth rates and migration analysis to determine age distribution of the population, and so forth.

For example, we can use expected industry productivity relations and the derived industry employments to estimate the income originating in commodity industries, while the income originating in noncommodity sectors can be estimated on the basis of state trends in the relation between total commodity and noncommodity incomes. However, the final projections of income originating must also be consistent with the final population estimates and the trends in state total and per capita personal incomes (which were observed in the first approximation model).

## *V. DETAILED PROCEDURE TO BE FOLLOWED IN OUR EMPLOYMENT ESTIMATES*

### *National Industry Totals*

The importance of national industry employment estimates to the regional projections requires that we describe briefly our technique for their projections. A complete set of national income and product accounts is projected by analysis of the demand and supply characteristics of the various sectors in the economy. This includes projected estimates of sales to final demand primarily on a 1-digit SIC basis. The industry final-demand sales are translated to industry gross output estimates (including intermediate sales) by using a projected ratio of final demand to gross output for each 1-digit industry. Changes in industry productivities are then projected and, finally, industry employments are derived by multiplying the reciprocal of the projected productivity by the industry gross output.

One-digit SIC detail is not sufficient for purposes of regional analysis. At least 2-digit detail within manufacturing is required. Several techniques will be followed in arriving at this detail as none of them can be considered wholly adequate by itself. It is hoped, however, that the use of alternative techniques will improve the judgments underlying our final results.

The first technique is to distribute the total manufacturing final demand sales on the basis of extrapolations of the proportions of gross product originating in each 2-digit industry within manufacturing. These value figures can then be translated to employment estimates through the use of productivity estimates.

A second technique is to project directly the final demand or gross output sales of 2-digit industries, and then derive the employment projections through productivity estimates. The projected 2-digit final demand or gross output sales can be based on extrapolated trends of the Federal Reserve Board market and production indexes, constant dollar shipments data, and constant dollar consumption and investment expenditures, all in their industry detail; or these data could be correlated with projections of "independent" factors such as population or some other industry's output; or, finally, their projected proportionate distribution within the manufacturing aggregate could be used as the basis for projections.

A third approach is to project changes in some of the critical input-output coefficients and then, using an aggregate form of the interindustry table and the projected final demands, to arrive at gross output projections which will lead to employment estimates.

It will be noticed that these various techniques for projecting national industry employments do not use employment trends directly but derive employment through sales estimates. The reason for this is that the national employment estimates should be consistent with the income and product account projections which themselves are interpreted as being sustainable and balanced pictures of the interrelations within the economy. Supplementing these techniques, however, will be analyses of national industry employment trends and specific industry studies in more detail.

## *Regional Distribution of National Employment Estimates*[9]

We now face the task of allocating each of our thirty national employment control totals among the states. Estimates of employment in the lead industries are considered first; they form the heart of our system.

### *The lead industry sectors*[10]

In the preceding sections we have outlined briefly the reasons for our choice of a three-stage trend-supply-demand approximation to lead industry employment projections.

FIRST APPROXIMATION. In the first approximation the trend approach will yield a $22 \times 50$ matrix of first approximation state industry employments. The twenty-two industry groups include agriculture, mining, and twenty 2-digit SIC manufacturing industries.

Our first approximation will extrapolate past trends in the state share of each national industry employment and distribute the 1975 projected national employments on this basis. These estimates

[9] Within the context of the national judgment model projections, we can introduce regional alternatives by drawing out the implications of how such factors as differing defense budgets might affect regional growth.

[10] Unless there is strong evidence to the contrary in specific instances, we will adopt the simplifying assumption that manufacturing, agriculture, and mining comprise the lead industry sector.

will then be adjusted to take into account trends in a state's share of total U.S. commodity employment and the possible decentralizing impact of an increasingly flexible transportation system. The first stage, therefore, not only by-passes the question of which are the causal determinants of employment shifts, but it assumes that the net effect of conflicting locational forces on the differential growth of regions will continue as in the past.

The first approximation will be broken down into three steps.

Each of the twenty-two state industry employments is first derived on the assumption that the 1957 state share of the national industry's employment will continue in 1975. Since each state had a different industry structure of employment at the 1957 starting point, and therefore a different mix of potentially fast- and slow-growth industries, faster rates of total employment growth will be displayed by those states whose 1957 structure favored the industries for which the NPA national estimates project rapid growth. This first step is useful mainly in introducing the proportional employment effect into our projections.

The second step introduces historical information on employment trends. Each state industry's share of the national industry employment is extrapolated on the basis of the evolution of its share between 1929, 1939, 1947, 1950, 1954, and 1957. Implicit in this step is the assumption that recent trends affecting state industry differential growth will continue without acceleration, deceleration or reversal.

The third step in the first approximation takes into account certain changing relationships that have a nationwide pattern.

These include, first, a comparison of the already derived total lead industry employment in each state with the trend in each state's share of total U.S. lead industry employment.

Second, the lead industry employments as derived will be used to obtain the first crude estimates of state lead industry incomes, from which total state incomes can then be derived. These estimates will then be compared with analyses of the trends in the state shares of national income. Significant discrepancies which appear from these comparisons could lead to adjustments of the industry employment projection.

Finally, at this stage we will introduce some of the expected effects of an increasingly flexible transportation system on the nationwide pattern of decentralization. For example, for some industries a declining ratio of raw material weight and cost to the value of final products has weakened the pull of raw-material sources and has enhanced the pull of the market in location decisions. In addition, the increasing importance of road transportation has reduced the costs of shipping in small lots, generally facilitating the establishment of smaller plants closer to the market.

These and other transportation developments are introduced at this early stage because they will affect differentially the various lead industries. For some industries, the changing structure of freight costs is already reflected in the present-day pattern of decentralization, so that a slow-down in the rate of decentralization of these industries might be expected; for other industries, the effect of recent transportation changes is still to be felt so that, for them, an acceleration of recent decentralization appears more likely.

*First Attempt at State Population and Total Employment Estimates.* Before we move to the second and third stages of our projections, it is important that we take one intermediate step necessary to a proper study of the impact of supply and demand elements on regional differential growth. This involves adequate projections of 1975 state population and total employment.

To derive state population, we will multiply total state lead industry employment by a projected ratio of the state's population to lead industry employment which can be derived from the trends in each state's ratio. In some cases the index of industrialization (ratio of manufacturing employment to state population) may be preferred as a basis for our calculations.

Total state employment will be obtained, at this stage, by deriving the various noncommodity industry employments from the trends in each state's share of the national industry employments and the trends in each state's ratio of employment to population.

SECOND APPROXIMATION. The second approximation is concerned with improving the lead industry employment projections by specific considerations of various supply and demand factors. The problem is, first, to determine for each industry the factors that are particularly important to its output and location decisions and how this importance might be expected to change in the future. For those industries in which a combination of the factors is important, it is necessary to provide weights for the various factors. Often this problem can be made simpler by disaggregating to a 3-digit level. The second problem is to obtain, partially on the basis of the first approximation, some idea of the ranking of states with regard to the various supply of these factors. This is necessary in order to convert the first approximation lead industry employment estimates by allowing for the effect of location of different factor endowments among competing states.

We have selected only a few of the supply factors which we consider of particular usefulness for projection purposes. The following illustrations should give some idea of how we plan to introduce the knowledge gained about the specific factors considered in order to convert the original trend-derived estimates.

*Relative Advantage in the Quality of Labor.* The rubber industry gives us an example of a case where relative advantage in the quality of labor supply is important. Once a dominant locational factor in the rubber industry, the quality of labor is losing much of its pull on industry location. In the past, access to a relatively large specialized labor force has been a major determinant of location in the rubber industries; the accident of first location in Ohio has forced much of the subsequent increase in employment in the industry to take place in that area. With mechanization and the reduced significance of highly specialized labor skills, the industry has been decentralizing rapidly.

For projection purposes, the important consideration is to determine whether the reduced importance of labor skills is already fully reflected in the present regional distribution of the industry, in which case no further decentralization need be projected on this account; or whether the working out of this factor will induce an accelerated, decelerated, or constant rate of decentralization.

Illustrating the increasing importance of quality of labor is the supply of high-quality technical and scientific personnel, which is likely to be an increasingly important factor in the "new" industries. For them, our research should yield at least some idea of whether proximity to intellectual centers (Massachusetts, the New York area, northern California) which have a large supply of only partly mobile trained personnel, is to exert an increasingly strong pull on location and account for differential growth in favor of these or similar areas.

*Relative Advantage in the Quantity of Labor Supply.* The secular decline in farm employment and the reluctance of some of the supernumerary farm population to move to large industrial centers are causing chronic agricultural labor surpluses in certain areas. The surplus labor is, in fact, available for industrial work, generally of a nature that requires little or no training. Such areas, both because of the size of the available labor supply and because of lower wages generally prevailing in agricultural communities, should be attractive to traditional industries such as textiles and leather. Identifying the areas of expected surplus agricultural labor supply in our first approximation study of state agricultural employments will give us useful information on surplus labor potentially available for industry in the various states.

*Relative Advantage in Natural Resources Availability.* Examples abound of situations where natural resources are the dominant locational factor. This is particularly obvious for those mining industries—such as copper—whose output is limited to a few regions but is used in manufacturing throughout the country. Examples can also be found in manufacturing. However, for many manufacturing industries, the secular increase in the value added by manufacture relative to the value of raw materials will make many processing industries relatively less dependent on raw materials location.

Natural resource availability is still the dominant factor for some manufacturing industries. For paper and allied products, for example, disaggregation to the 3-digit level shows the pulp and paper subgroup to be closely tied to lumber, its principal raw material input. Knowledge of future areas of lumber supply will therefore determine largely the future regional distribution of employment in the pulp and paper industry.

*Relative Advantage in Transportation Costs.* Transportation cost vulnerability will have a differential impact on the regional location of a given industry. Our study should reveal, for example, the extent to which a projected decline in freight costs to the Great Lakes area, resulting from the St. Lawrence Seaway, will affect the relocation of employment in the metal fabricating industries to that area.

*Relative Advantage in Locational Linkages.* The existence of strong forward or backward linkage effects can be revealing for the future location of a given industry, once a determination has been made of the location of the industry or industries closely associated with it.

One example of this situation is the tire and tube segment of the rubber industry whose locational shifts are closely related to shifts in the location of employment in the automobile industry. Future changes in regional employment in tires and tubes will therefore be highly correlated with a pattern of marked decentralization in automobile assembly, which is expected to continue in the future.

*Relative Advantage in Large-Scale External Economies.* The relevance of this consideration to regional analysis is that many regions of the country are gradually approaching a threshold size for well-diversified industrial structure. In the newly industrialized regions where development is already fairly advanced, such as California, additional external economies can be realized which are largely related to the ease of supply of semi-fabricated inputs from related industries.

We will be particularly concerned with those areas for which our first approximations of population, manufacturing employment, and other indicators suggest that the threshold stage may be reached by 1975. Florida might be one such example.

THIRD APPROXIMATION. Further consideration of the relative advantage offered by states in the size of various types of markets as they relate to industry needs, is the concern of the third approximation. The following illustrations will serve to show how we propose introducing this information into our projections.

*Final Markets in the State.* For some industries, the existence of large final markets in the state makes it possible to associate their employment shifts with changes in market access, which are given by projected state population (sometimes total state employment), as derived in our first approximation. One example is the stone, clay and glass products industry which prohibitive transport costs tie to the state market. The pull of raw material availability is not felt since stone and clay are virtually ubiquitous. For this and similar groups, a ranking of states in the order of their population gains as determined by our state population estimates will give us a strong indication of the states whose stone, clay, glass employment is likely to require adjustment from the first approximation.

*Final and Intermediate Markets in the Multi-State Regions.* Research is likely to reveal a number of industries where markets—final or intermediate—lie largely within the confines of a broad multi-state region. Ranking the regions in the order of their 1975 population (or total employment) as determined by our first approximation will point to the regions where our original industry projections require adjustment. Allocation of employment to individual states within the region will then be made largely on the basis of trend extrapolation of state shares of regional industry employment.

*Employment-Population Correlation.* Available data suggest that there is a high degree of rank correlation between total state population or employment and state employment in a number of industries, such as: primary metals; fabricated metals; machinery, except electrical; electrical machinery; transportation equipment; and instruments. A similar relationship has been found to exist for 3-digit groups such as apparel and related products, and printing and publishing. This information can be used as a check on our employment estimates as provided by our three-stage approximations.

### *A second look at lead industry and state population projections*

Up to this point, we have subjected each of twenty-two lead industry groups in each state to a trend-supply-demand analysis as adjusted by individualized estimates embodying a fair amount of state and industry research. It now becomes desirable to check these estimates against a series of independent expectations, based on observation of past time series and on our knowledge of interindustry relationships. Although some of these tests have been applied at earlier stages of analysis, they should be applied again in light of the improved employment estimates.

First, an attempt must be made to examine the total lead industry employments, state by state. Comparing each state's projected lead industry employment with the projected total United States lead industry employment gives, for each state, a projected share of total United States lead industry employment. That projected share can then be compared with a trend extrapolation of the state's historical share. Should these state shares appear to diverge noticeably, re-examination of the projections of individual industries will be necessary.

A similar check involves the calculation of the implicitly projected state indexes of industrialization (ratio of manufacturing employment to total population) which can be compared with trend extrapolations of state industrialization indexes, allowing for the steady tendency displayed by states to regress to the moving U.S. mean. This secular trend reflects a tendency for manufacturing to become an increasingly important percentage component of total activity in the new industrial areas such as the Far West, and a declining relative component of activity in the older industrial areas of the Northeast and the Middle West.

The internal structure of lead industry employment within each state should also be subject to scrutiny. Important locational linkages should be studied again, especially in states and in industries for which agglomeration economies are an important consideration. If expected linkages do not show up in the projections, then the projections should be reconsidered. For example, there is an expected correlation within states among durable industries, furniture, and apparel employments. Thus, a state which shows one of the highest rates of employment growth in, say, fabricated metal projects, will be expected to show a higher-than-average rate of employment growth in transportation equipment.

We have now completed all three stages of the lead industry employment estimates and checked them for historical and cross-section consistency. The regional distribution of employment obtained after the third approximation will be different from that of the first approximation, which provided the first set of state population projections. Therefore, second sets of state population estimates can now be derived from the third-stage lead industry employment projections. These new population estimates will help in projecting derived industry employments.

### *The derived industries*

Derived industries include the noncommodity industries unless there is strong evidence to the contrary in specific instances.[11]

Very rough employment estimates for the derived industries were obtained in the first approximation on the basis of trends in the state share of each national derived industry employment. We are now prepared to replace these estimates by projections based on analyses of market demand factors.

The most important markets for derived industry outputs are the final and intermediate markets within the state. The best indexes of the size of these markets available to us at this stage are projected state population and commodity employment. The historical data indicate that the share of total state employment devoted to most of the noncommodity industries (with exceptions such as finance and the federal government) does not generally vary widely among states, and that even these differences have tended to decrease as the state ratios are regressing to the moving U.S. mean.

[11] The noncommodity industries specifically are communications, transportation, trade, business, professional, repair and personal services, finance and insurance, real estate and rental, private households and nonprofit institutions, state and local governments, federal government, and construction.

Thus derived industry employments can be obtained by extrapolating, for each state, past ratios of specific noncommodity industry employment to total lead industry employment and to population. These ratios will implicitly incorporate the correlation effects of past changes in the size of markets outside the state.

Supporting this analysis should be an examination of the effects of regionally different income distributions and how they affect household expenditure patterns. It is also necessary to analyze separately those noncommodity industries such as the federal government, and the finance industry in some states, which are not particularly related to final or intermediate markets within the state.

## *VI. REGIONAL PROJECTIONS OF OTHER INDICATORS*

As we have indicated earlier, regional projections of indicators other than employment will be derived from the regional employment projections whenever possible. Some of these indicators have already been mentioned as building blocks in the procedure for projecting industry employments. This indicates not only that preliminary projections for these indicators will often be necessary at earlier stages, but also the mutual interaction among the various indicators. It implies that still further improvements in the employment projections can be obtained by reviewing them in the light of these other indicator projections. This last section provides a brief description of some of the problems expected to arise in making the projections.

Population, already projected in the aggregate for individual states, should also be distributed by age groups. For most states we will expect a rough correspondence with the national average. Some farm states, however, will show a different age distribution because of rural-urban differences in birth rates and because of farm migration. In addition, some of the resource amenity states are likely to be particularly attractive to the aged group in the population. An urban-rural distribution of the projected population should prove useful and might be based on the industry employment characteristics already derived as well as assumptions about the extent to which rural areas will attract industry.

The labor force in each state will tend to show the average national relationship to employment, except in some specific instances where depressed industries are an important part of the state industrial structure.

The various state income estimates should not only be consistent with the state employment estimates but with each other as well. State personal income can be estimated by extrapolating the trends in state personal incomes, state shares of U.S. personal income, and state per capita incomes.

Value added created and employee compensation paid by each state lead industry can be obtained by use of the industry employment estimates and expected productivity changes. The productivity changes should take account not only of over-all increases for each national industry, but also current regional differences in productivity for the same industry as well as any expected lessening of these differences. Decreasing regional differentials in value added or employee compensation per worker are likely to be significant for only a few states, since the historical data show that high-growth, low-income states' incomes per worker in manufacturing have grown at a slower rate than total state incomes, suggesting that the state industry growth rates are comparable to those of the national industry. Agriculture productivity, on the other hand, appears to be rising faster in the low-income agricultural areas than for the nation as a whole. State-derived industry projections of value added and employee compensation, in addition to working through productivity changes, should take account of the historical observation that for individual services these incomes vary in response to changes in the relative position of the state on the U.S. per capita income ladder.

State personal income can be again derived by adding specific other earnings to employee compensation. Federal government transfers to each state can be derived by a per capita allocation to each state, except in those instances where population and unemployment estimates suggest strong departures. State government transfers will depend on state-by-state analyses. State proprietor incomes can be derived by relating them to the state service industry employment estimates. Rental income, dividends, and interest earnings will each require special analyses and can probably be related to state population. Special studies will also have to be made in those few areas where interstate commuting makes for a significant difference between incomes earned by state residents and incomes paid out by the state industries. The state personal incomes derived by this component build-up method should be compared and made compatible with the income projections previously obtained by the methods of allocating U.S. personal income.

State household income distribution projections will tend to conform with the logistic curve of national income distribution. Some state adjustments in the shape of this curve will be necessary in order to take account of, primarily, the differences between farm and nonfarm households. Since average state incomes will be different, state income distributions will also be different even when the logistic curve is the same.

State lead industry sales can be projected by relating gross output to industry employment or value added. The state distribution of personal expenditures can be projected by adjusting the national distribution on the basis of information on state differences found in consumer expenditure studies.

In concluding, it might be well to emphasize again that our choice of indicators to project is based primarily on their potential for immediate utility in public and private policy decisions. In addition, we believe that they are essential to regional accounts construction, although their final utility in this area awaits further experimentation with regional accounts.

## COMMENT

*Gerhard Colm*

*National Planning Association*

The approach recommended by Sidney Sonenblum and Louis Delwart is much more limited than the recommendations for local or regional accounts made by other speakers. They were prudent not to promise a full system of local or regional accounts for early delivery, but rather to suggest a few key indicators for the various regions. Such key indicators may be of some direct use for private and public decision making; they can also serve as guides and checks for those who are concerned with the construction of local economic accounts.

Sonenblum has proposed basically three steps. One is based on the present product mix in various regions and assumes a faster rate of growth in regions now having a larger percentage of the growth industries. A second assumes a slower rate of growth in regions now having a large proportion of lagging industries. This approach requires the processing of a considerable amount of statistics but poses no difficulties in principle. The third step, also, namely the adjustment to the demand side of regional developments, poses no difficulties in principle.

The second step, however, requires adjustments to locational factors. These adjustments may involve, first, industries which utilize at a future time locational advantages that are now already in existence. That may lead to a different product mix in the region than we have now. Second, new locational advantages may be developing. For instance, there may be migration of labor to new areas which appear attractive as a place to live; then industry may follow the movement of labor, which is the opposite dynamic causation from the one assumed in the first step. Sometimes industries may only discover a locational advantage which already exists. Third, entirely new technological developments may affect the locational factors. Finally, this differential effect may also result from deliberate national or local policy. Here we have to rely on our judgment as to what extent such policies will be adopted and effective.

In conclusion, let me raise a question concerning the federal statistics program. I remember that the National Accounts Review Committee, under the chairmanship of Ray Goldsmith, assigned a rather low priority to a regional breakdown of national economic accounts. The participants in this conference seem to give high priority to the development of data that would permit regional economic accounts. I also am under the impression that many participants in this conference feel that such data should be developed by the federal government rather than by surveys at the local level. Two proposals have been made, namely, (1) that many more data are needed, and (2) that statistical data should give less aggregates and more cross-classifications in order to show the interaction of various economic characteristics of the economic units. In connection with work undertaken by the Federal Statistics Users' Conference, it would be of interest to have recommendations concerning the desirability of improvements in federal statistics to meet the needs of private and public decision makers on the regional or local level.

## *Joseph L. Fisher*

*Resources for the Future, Inc.*

The paper opens with a few well-chosen paragraphs on the uses of regional accounts which, the authors are convinced, are principally for improvement of public and private policy. Understanding of regions for its own sake becomes instrumental only—a way station on the line to a better world. I point out this policy emphasis, not to disagree with it particularly, but to make clear that the authors put us on warning that they will select data and methods with this as the ultimate objective. The effort from the beginning is to be made operational; the investigators will not be afraid of exercising their good judgment when that is necessary to get past data and methodological roadblocks.

What theory or theories of regional economic growth underlie this exercise in practical methodology? Need there be any theory at all behind the projection of regional accounts? I am not sure about the answer to the second question although at the present stage of development of regional projections, the "selective art" stage, I am inclined to think there has to be a theory in mind. Regardless of this, I find implicit in the National Planning Association method certain theoretical notions beyond those noted explicitly by the authors. For example, the separating of "lead" and "derived" industries seems to be impelled by the idea that changes in "lead" industries, and the supply and demand factors underlying these changes, are the cause, or close to the cause, of differences in the growth of various regions. "Lead" industries are much the same as what others have called "export" or "autonomous" industries, or, more loosely, "economic base" industries. Also I found myself wondering what theories the authors had in mind about the role of investment in regional economic growth. In several places the authors state their belief "that regional economic changes are basically reactions to national developments." In some general sense this is true, I would agree, but how the region reacts, which region reacts, and to what extent the region reacts are determined regionally as well. Frequently, these are precisely the questions of paramount interest, answers to which cannot be read out of a book on national developments. The difference between the authors and me here is one of degree only, but *vive la différence!*

The assertion is made from time to time that regional economic differences are tending to disappear, or at least be considerably reduced, and that transportation and other improvements lead to more and more decentralization. There may indeed be something analogous to one of the basic laws of thermodynamics at work here with regional differences running to a constant, but I am not persuaded one way or the other by the evidence. Compared to colonial times, centralization of economic activities and regional differences in income may be much greater now. More typical of our time is greater centralization in larger metropolitan areas within which decentralization to suburbs is taking place. Transportation improvements, depending on their nature, may encourage centralization or decentralization. If there is a regression over time toward a moving mean, then the questions arise: regression of what economic items, over how long a time, and for what areas?

## *Douglas Greenwald*

*McGraw-Hill Publishing Co.*

As a business economist working for a profit-making company, I am looking for forecasts of fifty state economies expressed in real gross product terms or real personal income terms or industrial production terms. However, as a business statistician I know the difficulties involved in producing historical data, let alone reliable economic forecasts, on a state basis. Nevertheless, this conference must face up to the importance of economic projections as well as of the policy-making decisions which would follow.

In the future regional figures may prove to be far more important in economic analysis and prediction than national figures. Discussions with businessmen about industrial growth have made this clear to me. However, I am surprised that among this group of regionally oriented economists there isn't more agreement about the growing importance of subnational totals. For example, Delwart and Sonenblum state that regional development is *primarily* determined by growth and changes in the national environment. In other words, the national forecast must come first; the regional forecasts later.

Although the authors list thirteen critical indicators, they devote most of their paper to the technique of forecasting regional employment. They spell out in detail three estimating stages or approximations of employment and income. Their first step in deriving regional employment forecasts is based on the familiar shift and share approach. Next they go through various stages of improvements and refinements. When they are finished with their judgments on improvements and refinements, the so-called "improved" data no longer bear any resemblance to the figures derived from the shift and share stage.

New and significant developments are taking place right now which suggest major structural changes may be just around the corner. Thus the future course of economic growth—regional as well as national—may be affected by striking new developments. The authors of the paper do not give enough weight to these new factors in their projection techniques.

The first new development is research. Industrial laboratories throughout the nation are sending forth a stream of new processes, new techniques, and new products which may soon turn into a flood. It is my educated guestimate that about 30 per cent of 1970 sales will be in new products not yet produced. Therefore, forecasting national as well as regional employment, and other economic factors, based on historical data may turn out to be a somewhat meaningless exercise.

Another development which is beginning to take on significant proportions is the export of scientific and managerial personnel. I should add an even more important shift of recent years. This is the recent burst of investment of United States companies in overseas plants. Obviously, this development will have a downward influence on United States demand for employment in the near future.

Another significant factor is the development and general acceptance of the computer. No doubt the computer will cut significantly future demand for middle management personnel in business and government. It seems to me then that research and development, the export of plant and personnel, and the general acceptance of the computer may become so important in determining growth prospects of the future that we should just take off from some current base and estimate where we are going from there.

I have quarreled with only a few things the authors have said. Let me repeat they have done a fine job of spelling out in detail the techniques of a regional forecasting project.

# 8

# MICROANALYTIC MODELS FOR REGIONAL ANALYSIS

*Guy H. Orcutt*[1]

*University of Wisconsin*

THE PRIMARY purpose of this paper is to bring microanalytic models to your attention as potentially useful research and policy guidance tools in the area of regional analysis. A full account of this approach to the study and representation of extensive and complex social systems will be found in a forthcoming book, *Microanalysis of Socioeconomic Systems—A Simulation Study.*[2] This paper draws heavily on Part I of this book and on suggestions made by John Korbel.

The new type of model outlined in this paper has been developed with the notion of applying it at the national level, but it is just as applicable to regional analysis. In addition to describing microanalytic models and their solution by simulation techniques, this paper attempts to draw attention to some of the data problems to be faced in constructing regional microanalytic models.

The major advantages of microanalytic models, solved by simulation methods and constructed and used with the assistance of continuous data collection by sample survey techniques, come about by virtue of the following aspects of such models.

[1] I would like to express my appreciation for the several valuable comments and criticisms made by Professor Martin Beckmann and Dr. James W. Petersen in their roles as discussants. At various points I have tried to indicate specific criticisms of a discussant.

[2] Since published. G. H. Orcutt, M. Greenberger, J. Korbel, and A. M. Rivlin, *Microanalysis of Socioeconomic Systems—A Simulation Study* (New York: Harper and Brothers, 1961).

1) Decision units and other microcomponents may be used. This is highly important because: such units are readily recognizable units that can be observed rather directly and may be studied intensively on a continuing basis with data collection geared to research needs; there exists a substantial and rapidly growing body of knowledge about such units; the various activities of a decision unit are closely and to some extent simultaneously interdependent; policy actions of interest are implemented by actions that deal directly at the microlevel of decision units; and finally, because the well-being of such units and the manner in which they are affected by federal, state, and local policy is of central concern.

2) As many variables as desired may be used, and these variables may relate to aspects of microcomponents which are either difficult to aggregate or which lose much of their significance when aggregated.[3] Such variables would include locational and attitudinal variables along with others relating to education, birth, stocks of goods, and so on.

3) Since this approach does not require aggregation of microrelations, research workers have nearly complete freedom in specifying the operating characteristics of decision units and of other components.[4]

[3] Dr. Petersen commented that, "It is not true as Mr. Orcutt indicated that in a microanalytic model as many variables as desired can be used; since, first, we usually do not know enough to specify relevant relationships with any real confidence and, second, the introduction of many variables increases research costs at a frightening rate. This latter point almost always limits our endeavors."

I agree with the substance of this comment. My intention only was to indicate that microanalytic models, solved by simulation methods, permit the use of as many variables as desired insofar as handling and solving such models is concerned. The fundamental problem of discovering appropriate behavioral relations remains and, as Dr. Petersen comments, "Technique alone will not solve our problems although, as one reads the journals, economists at times seem to forget this fact. To build a useful model one still needs materials in the form of valid inputs, and the slow and tedious gathering of such inputs is our biggest need today as it will be tomorrow."

[4] Dr. Petersen comments that, "To take on any significant problem will require the aggregation of many microrelationships. The simplifications implied in such aggregation mean that the simulation is, as it always must be in economics, an abstraction of reality. This means, in turn, that all participants in the project must be aware of the relevant abstractions, and hence they do

4) This approach facilitates and encourages the use of a building-block approach to model construction.

The above advantages are as significant for regional analysis as for national analysis. Perhaps they are even more significant, since regional policy formulation must concern itself with rather specific and detailed matters if it is to be of maximum value. Microanalytic models excel in their capacity for effectively using detailed information and considerations. The major objective of incorporating such detail has been that of improving predictions about aggregates. This would remain perhaps the major justification even in applications at a regional level. However, the ability to incorporate detail, and the potential ability of yielding fairly detailed results, assume a larger significance at the regional level than they did in thinking about applications at the national level.

## *I. NEEDS OF POLICY-MAKERS*

Policy-makers in all areas have certain needs in common. They must make choices, and it seems reasonable to expect that such choices be made with objectives in mind. If their choices are to modify subsequent developments, there must be instruments that can be manipulated. If policy-makers are to specify the use of instruments, they must be able to recognize when the actual deviates from the desired. Given that they know what is desired, this at least implies the necessity of some knowledge of current status or position as it develops over time.

Objectives, instruments, and knowledge of current status are thus recognized as needs common to all policy-makers, but it also will be perceived that they are not sufficient. Policy-makers must have models which respond to the use of policy instruments in

---

not, as Mr. Orcutt suggests, have 'nearly complete freedom in specifying the operating characteristics of decision units and other components.' "

I agree, of course, that all models are abstractions no matter how micro or how detailed. The point I wished to make was that the use of a microanalytic framework and of simulation techniques can free the individual research worker from the necessity of using simplifications, such as linearity, in order to facilitate or make possible the aggregation of microrelations. He is enabled to choose among alternative abstractions on the basis of their apparent explanatory and predictive merits.

approximately the same way that the real system would respond. These models may be stated verbally, they may be mechanical contrivances, they may be graphical devices, they may be sets of mathematical relations, or they may be computer representations. The form is a matter of convenience, but the use of models is a prerequisite to intelligent policy formulation and execution.

The use of models in forecasting is of direct relevance to policy formulation but sufficiently evident to require little elaboration. The present situation is only known with a lag and instruments only work with a lag. Therefore if policies are to be appropriate, it is desirable that policy-makers be forewarned of impending developments. Accurate models of the solar system are of value in predicting the relative positions of the sun, moon, and planets, and such forecasting has an obvious significance for navigators. Population and labor force forecasts also are based on models and have an obvious importance to regional planners.

Experimentation with models of aircraft in a wind tunnel illustrates the case in which models may be useful in choosing between alternative designs. In this case, experimentation on a given model may consist of varying the air stream impinging on the model, introduction of turbulence into the air stream, operation of the rudder and other controls of the model, and so forth. It is hoped that the model responds to experimentation in the same way that the plane modeled would respond. If this is in fact the case, then this sort of simulation makes it possible to learn how various designs would work out in practice without the enormous expense in life and property involved in experimenting on real aircraft. Such conditional prediction does not eliminate the need for test pilots and experimental aircraft, but it does reduce the number of false tries with real aircraft and pilots.

Conditional predictions based on experimentation with a model of a real economy might be of assistance in choosing between alternative governmental policies relating to taxation, spending, etc., in much the same way that conditional predictions, about what would happen if such and such were done, are valuable to the policy-maker trying to select a plane design.

If policy-makers are considering one tax law as compared to some other tax law, or one spending policy as compared to some other spending policy, they ought to know what kind of consequences would follow from the alternative policies. The quickest way to find out the consequences might be to take a real economic system and try out the alternative policies. But it would not be very satisfactory from the standpoint of the participants in the system. We ought to choose between policies on the basis of their fruit, but it would be preferable to sample the fruit before trying out policies on a real economy. If a model of the economy were available it could be used to generate the consequenecs that would follow from application of the alternative policies. If the representation of the way in which the components of the economy respond and interact were incorrect, then incorrect conditional predictions would result. However, if the model were sufficiently accurate, such conditional predictions would be of enormous value in policy formulation and selection.

## *II. ALTERNATIVE FRAMEWORKS*

The preceding section has mentioned the common needs of policy-makers and has emphasized their need for accurate models. The purpose of this section is to present the alternative choices that must be made in building or choosing a model suitable for regional analysis and policy formulation.

The elements of a model of an economy are components, variables, and relations. The alternative choices open to the model builder have to do with choice of components, selection of variables, and specification of relations. It is convenient to group models into major categories, but it should be recognized that models may differ from one another somewhat independently in each of many dimensions.

Perhaps the most significant way in which models may differ is with respect to choice of components. Variables and relations relate to components so that if two models differ with respect to components they are certain to differ in a great many other respects.

In addition, data requirements and appropriate research strategies may be very different.

The different major types of components commonly recognized by economists in real economies are major sectors, such as the household sector or government sector; industries; and microcomponents, such as individuals, families, firms and specific markets. Models of economic systems, whether regional or national, could be built in terms of representations of any of these kinds of components or in terms of various combinations of them. Examples of models using major sectors as components would include models of economic systems constructed by Jan Tinbergen, Colin Clark, Klein and Goldberger, J. S. Duesenberry, and Arthur Smithies, to name only a few. Models of economic systems using industries as basic components primarily include those built by W. W. Leontief, Walter Isard, James Henderson, Leon Moses, Werner Hirsch and several others. Some of these models classify industries on a regional basis. Models of economies built in terms of microcomponents are still in a formative stage. At a conceptual level the Walrasian general equilibrium model would fit into this category.

Even if two models were built in terms of identical components, the variables used in describing the behavior of the components might still differ in many important respects. Part of the difference might reflect a difference in interest and in part it might reflect a difference in understanding of the behavior of the components. If a model is to be used for policy purposes, it seems obvious that the variables in it should include some kind of representation of each of the instrumental variables that might be used in executing policy, and that it should generate those variables needed to describe adequately the behavior of the regional economy in question.

Even using the same framework of components and variables, models may differ substantially with respect to the relations serving to link current endogenous variables to exogenous and lagged endogenous variables inclusive of policy instrumental variables. Equilibrium models may be built up out of relationships that do not involve use of any time lags, whereas dynamic models will involve time lags or derivatives of integrals. Conceptually, at least, models built in terms of any kind of components and with nearly any kind of variables could be either equilibrium or dynamic in nature.

Another fundamental way in which relationships might be different is whether they are functional or probabilistic in nature. Either might be the case in models composed of any kind of component and using any kind of variables.

Models also differ with respect to form of relations used. Many available models use relations which are linear in the variables, and nearly all available models are linear with respect to parameters. It does not seem likely that such broad usage of linearity assumptions in either variables or parameters can be justified on any grounds except that of expediency in the face of inadequate tools of estimation and analysis and in the face of inadequate data.

## *III. MICROANALYTIC MODELS*

The previous section described several ways in which models of regional economies might differ from each other. This section sketches out in somewhat greater detail a microanalytic framework for models of regional economies.

A distinctive characteristic of a microanalytic model is that it contains components corresponding to microcomponents of real economies. These components fall into three broad categories: decision units, markets, and goods.

In an economy it is possible to identify several different types of decision units—individuals, families, firms, banks, labor unions, local governments, etc. A microanalytic model of the economy would contain a population of decision units composed of a relatively small number of different types of decision units and a relatively large number of units of each type.

In the economy it is possible to identify many different markets. A microanalytic model of the economy would contain many different markets and these components would provide linkages between the decision units in the model just as real markets serve to link potential buyers and sellers or borrowers and lenders.

In the economy it is possible to identify many different types of goods, services, credit instruments, and other things which are used or held or enter into transactions. A microanalytic model like the economy would contain a constantly evolving population of these things which are used or held or produced by decision units and which pass through markets on their way between decision units.

Variables in a model relate in one way or another to the components of the model, and the name of a variable should be explicit enough to make clear the component or components to which it relates. It is convenient to classify variables used in microanalytic models into three broad categories: output variables, status variables, and input variables.

The decision units have various possible behavior "outputs." Individuals may go to school, enter the labor force, earn wages, get married; married couples may set up new households, have children, get divorced; firms may purchase raw materials, hire labor, establish new plants, produce goods, and so forth. Any observable behavior of a decision unit—even the expression of an opinion or attitude—could be considered an output.

The outputs of a decision unit during a given time period are taken to depend on prior inputs to the component and on the values of the component's status variables as of the beginning of the period.

Anything external to a decision unit which acts on it or influences its behavior may be considered an "input" to the component. Inputs may thus include the seasons, the weather, time, and the prior outputs of other components.

A component's status variables are internal variables which describe the state of the component. The values of these variables as of the beginning of a time period may influence the behavior of the decision unit during the time period, and the output of a decision unit during a period may alter the values of the status variables that will describe the state of the decision component as of the start of the following period. Status variables are characteristics of the units themselves—initially assigned in the proportions in which these characteristics appear in the real system. For example, status variables of individuals might include age, sex, marital status, education, income; status variables of firms might include inventories, sales in the last quarter, back orders, balance sheet variables, even anticipated sales.

In addition to components, and the variables relating to the components, a model of an economy must also contain relationships if it is to generate any predictions. Relationships specify how the values of different variables in the model are related to each other or how they are otherwise generated. Relationships are of two broad types: identities and operating characteristics.

Identities are accounting or tautological statements that may be introduced for convenience. Thus total sales of a product are set equal to total purchases, total assets of a firm are set equal to total liabilities and equity, etc.

An operating characteristic is a relationship specific to a given component which specifies either an hypothesis or an assumption about how an output variable of the component is related to status variables and input variables of the component. Operating characteristics, by specifying how components respond to stimuli or how they generate outputs even in the absence of external stimuli, embody much of the real knowledge brought to bear, and it is primarily upon them and the initial distribution of status variables that any predictive use of the model must ultimately depend. And, whereas components and variables relating to components may be directly observed in the real system, it is important to recognize that operating characteristics cannot be directly observed but rather must be inferred by inductive methods of analysis. Operating characteristics may be of any form research indicates to be appropriate.

In general, the relations linking outputs of a decision unit to the prior inputs to the unit and to the status variables of the unit are considered to be stochastic in nature; i.e., it is the probabilities of occurrence of certain outputs, rather than the outputs themselves, which are regarded as functions of the input and the status variables. The probability functions which relate outputs to inputs and status variables are called the operating characteristics of the component. If the probability that an individual will marry in a given period is taken to depend on his age, sex, and marital status, then the table (or function) which specifies the probabilities of marriage for males and females of various ages and marital conditions would be one of the operating characteristics of the individual. This probabilistic approach to predicting behavior does not reflect a philosophical position about the nature of causation. It simply represents an attempt to utilize the kind of information which we actually have about the decision units of our economic system.

One advantage of describing a model in terms of components, variables relating to the components, and relationships between the variables is that it facilitates a building-block approach to the construction and testing out of the model. The components serve rather naturally as the major building blocks and facilitate a useful subdivision of the over-all effort required to achieve a useful model of an entire economy. The building of models of each kind of component may then proceed with only a very limited co-ordination required between the groups working on different components. Each research group needs to be aware of what outputs are being generated by other components and what outputs are expected from its kind of component, but that is all. Each is free to be guided solely by the evidence its research turns up in generating the outputs expected from its components.

It is now necessary to elucidate the matter of input-output linkage between decision units in the model. Some outputs are simply end-products and nothing more is done with them except to aggregate them and read them out as desired. Most outputs, however, update status variables of the same unit and/or become inputs to other units. As it will be recalled, the second major class of components in microanalytic models are markets, and the specific function of these components is to transmit the outputs of decision units and to distribute them as inputs to other decision units.

Individual decision units that wish to sell something make known the prices and terms under which they are prepared to sell. These outputs of decision units are fed in as inputs into appropriate markets. The operating characteristics of a given market summarize the offers to sell that it receives. The summarized information about price, location of seller, delivery dates, etc., is then used in other operating characteristics of the market, and the outputs of these operating characteristics are fed to potential buyers. The information fed to a potential buyer can be made to depend upon the potential buyer's location relative to that of the seller.

The information received by potential buyers is used in their operating characteristics, and in some cases orders to buy are the resulting outputs. These orders enter as inputs into the appropriate markets. Operating characteristics of each market summarize and classify the orders by region of origin, price accepted, etc. Other operating characteristics of the market then use this summarized information about orders, along with information about each potential seller considered in turn, to distribute the orders among the potential sellers. Decision units respond to the orders which they receive as inputs and generate various outputs among which in due course usually will be deliveries. These deliveries show up as inputs into an appropriate market, are transmitted and distributed by the market, appear as outputs of the market, and as inputs into the appropriate decision units. These decision units then make payments or promises to pay which are again distributed through the market to the appropriate sellers, and so forth.

Figure 1 shows graphically how the outputs of the entire set of *N* decision units are transmitted and distributed by the *M* markets. The outputs of the decision units become inputs into the markets. After being summarized and distributed they appear as outputs of the markets. These market outputs then become inputs into the decision units again.

An essential characteristic of this type of model is that it proceeds in short discrete steps or periods. It is thus a recursive model in that the outputs of a unit in any period depend on prior inputs to the unit. Since there is no simultaneous interaction between units, there are no simultaneous equations involving more than one unit at a time to be solved. This does not mean that units are conceived of as acting independently of each other, since the prior

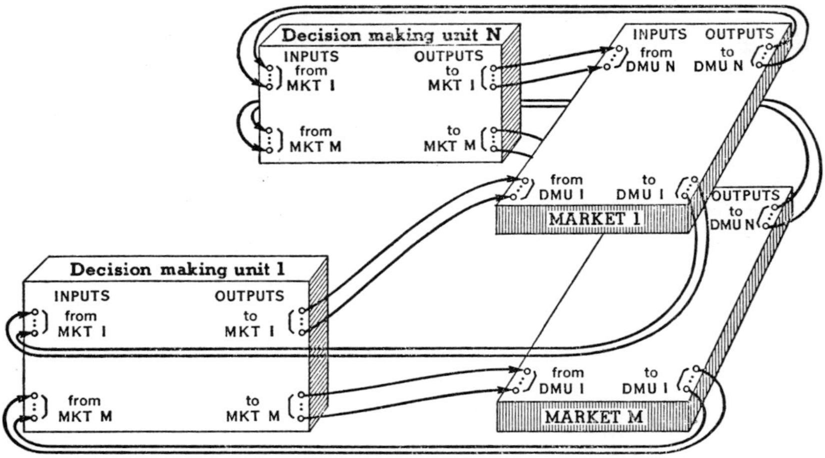


*Figure 1.* Incomplete flow diagram of an economic system.

outputs of other units may be inputs to the unit in question; but it does mean that all interaction among units in the model is sequential rather than simultaneous. However, individual components may contain simultaneous equations if this is desired.

If a recursive model is to be a good approximation to reality, the time period chosen must be short. Many responses, such as the response of purchase probabilities to a change in price or the reaction of marriage probabilities to a change in income, may be sequential in fact, but will appear simultaneous if annual or other long-period data are analyzed. We have used the month as our basic period. For some purposes an even shorter period, such as the week, might be preferable.

The implications of a microanalytic model are worked out by simulation on a large electronic computing machine. As a first step, the components are assigned initial values of their status variables in the proportions in which the corresponding characteristics appear in the components of the real economy. These units with their status variables are listed in some arbitrary order on a magnetic tape which is fed into a computer. The simulation proceeds one period at a time. In each time period each individual unit is considered in turn. For each possible output of the unit, a probability of occurrence is specified by the relevant operating characteristics that are used in connection with the appropriate status variables and inputs to the unit. Whether the output occurs or not is determined by a random drawing from this probability distribution. For example, suppose the simulation is started with the month of January, 1960, and the first unit considered is an individual with the following characteristics (among others): male, white, aged 34. We have already estimated the probability that an individual with these characteristics will die in this month at, say, .0002. Let us suppose that .0002 is the probability of death specified by the relevant operating characteristic. Now a random drawing is made from a distribution using random numbers generated by the machine. There are two chances in 10,000 that the outcome of the draw will indicate "death" for this unit and 9,998 that it will not. The man either "dies" and is eliminated from the population (and from the household or other larger unit of which he may be a member) or he "lives" through the month—depending on the outcome of the draw. There may, of course, be more than two possible outcomes. If the unit were a household, for example, the output "amount spent on durables" might have several possible values—e.g., $0, $1–100, $101–200, etc.

When each possible output for each unit has been considered in this way, the first "pass" or month is complete. We enter the second month with a population of units which is slightly different both in size and composition from the initial one. Some individuals have died or married, some babies have been born, some firms or households have been created or destroyed, many have altered their asset and debt positions, and so forth. The whole procedure is then repeated for the second month and for as many more as desired. To find out what has happened to the system at the end of, say, twelve months, we take a "census" and actually count the number of units with various characteristics, or combinations of characteristics.

In a model involving household units, firms, financial institutions, etc., there would also be many markets. The operating characteristics of these markets and status variables relating to them would be stored in high-speed storage. As decision units place orders, for example, these orders would be summarized by operating characteristics of the appropriate market and the summarized information retained in high-speed storage. Then these orders would be distributed to the appropriate sellers as these decision units pass through the computer.

## *IV. DATA REQUIREMENTS*

To gain a better understanding of the way in which components such as households and firms respond, behave, and interact, requires intensive scientific research—both co-operative and individual—by scientists from many disciplines. The future direction which this research will take may be difficult to chart, but some features about data requirements are fairly evident.

To learn about the operating characteristics of any component or unit, one would like to observe the behavioral or dependent variables of interest and to vary possible stimuli, or what I shall refer to as explanatory variables, first one at a time, in a variety of time patterns, and then in various combinations. Given enough such experiments, it should be possible to discover the important explanatory variables, as well as how their variations over time are related to the variations over time of the dependent variables.

Insofar as experiments of the above type are feasible, we should carry them out. However, given the nature of the units, the size of potential research resources, and the limitations imposed by our ethical beliefs, it seems clear that only a small part of the needed knowledge will be gained from planned experimentation.

Instead, we must make use of experiments which are conducted for us by the economic system itself. We must search out situations in which possible explanatory variables vary in something like the way we would have liked them to vary in an experiment. For each unit studied we must observe these possible explanatory variables and we also must observe the behavior or response of the same units.

Since we are not entirely sure which variables are relevant or how they are relevant, we must be able to measure a wide variety of possible explanatory variables for each unit studied. Furthermore, since responses are likely to lag behind the explanatory variables, we need to measure, at different points in time, a wide variety of variables relating to each unit studied.

One other data requirement is that it must be possible to adapt rather quickly the collection of data to changing research requirements. Hypotheses must be formulated; data appropriate for testing must be collected and evaluated; hypotheses must then be modified or new hypotheses introduced; data appropriate for testing these hypotheses must be obtained; and so on. This continual cycle of hypothesis formation and testing is common to all scientific research, and demands a data collection instrument which is highly responsive to changing research needs.

Since experimental techniques are largely out of the question, we are left with the alternative of using sample survey methods to obtain data which are needed, or of relying primarily on use of data which are collected for administrative purposes by governmental agencies. With certain significant exceptions, data now collected and made available by governmental agencies for research purposes are of limited usefulness for the research which is needed on the behavior of households and firms and other component parts of our economic system.

One reason for this is that data are frequently collected, processed, and made available in such a way that almost all connection is lost with households and firms identifiable over time. Instead, aggregate figures are sought and these are pieced together from many sources. For example, it may be that from the national income data accurate measures are obtained of aggregate income. It also may be that accurate measures are obtained of aggregate consumption expenditures classified by type of good. But there is no way of using these data to test hypotheses about the relation of household consumption expenditures to household incomes except by incorrectly assuming that all households are homogeneous in their response to income.

And if one wishes to test hypotheses about the influence exerted upon consumption by children, existing stock of durables, debts, monetary assets, home ownership, income change, interval since marriage, attitudes, intentions, expectations, location, age, retirement, social background and environment, advertisement, occupation, source of income, capital gains, etc., one is reduced to the absurdity of obtaining aggregate United States time series on these variables and then seeking to test something by relating those aggregate time series to aggregate consumption figures.

The position we are left in, then, is that to get a more useful model of a region we need to investigate the behavior and interaction of its fundamental components. To do this, we need to be able to measure a substantial number of input and output variables at successive time periods for each of a large number of households and firms. The only known way of getting such data for this kind of unit is by survey techniques.

Furthermore, we must be able to collect data in quick response to changing research needs and do this at a cost which can be met. This means that we must draw our information from scientifically selected samples. In other words, sample survey statistics are indispensable to effective research of the type needed.

Construction and use of microanalytic models at either the regional or national levels calls for some reorientation of our thinking and effort with regard to data collection and use. However, essentially the same reorientation is called for in either case, and it is high time that data collection be linked more effectively to basic research needs. Theory making, testing, and model building should guide data collection.

We need to move toward a co-ordinated system of data collection from individuals, households, financial institutions, firms, and local, state, and federal governments. Cost considerations dictate data collection on a sample basis. Research considerations and possibly cost considerations suggest that data should be collected from geographic clusters of the above decision-making units. Clustered samples are frequently used, of course, to reduce interviewing costs. However, at present inadequate attention is given to achieving appropriate overlaps of sampled clusters of different, but interacting, types of decision units.

From a research standpoint, we are interested in behavior over time as well as in interaction. This suggests the importance of obtaining re-interview data or, better still, data over time from rotating panels.

The data that are collected need to be processed and made available for research purposes in a form which avoids aggregation of data relating to different decision units. When desirable, the research worker can aggregate disaggregated data. The reverse operation is not possible because the essence of aggregation is the discarding of detail about individual units. Effective research would not seem to require identification of individual respondents, but it does require that data about microcomponents be processed and made available without either severing or averaging out connections that exist between the separate pieces of data.

## COMMENT

*Martin J. Beckmann*

*Brown University*

Mr. Orcutt is to be congratulated on an excellent paper. He has set before us with admirable clarity the principles of model building in the social sciences. He has presented a persuasive argument for "microanalysis" wherever feasible. Above all, he has drawn our attention to a novel and versatile tool: simulation—a method which allows microanalysis to succeed where intuitive or mathematical analysis would bog down. He would argue that in regional studies we are perhaps not sufficiently aware of the vast possibilities opened up by the computer revolution. On the other hand, a certain "wait and see" attitude on our part may be excused until some of the current important and ambitious experiments with simulation methods have been completed.

On the broad methodological points raised in this paper there can be very little disagreement. My discussion can thus be very brief. In fact it consists of no more than four questions for further information:

1. While this paper is eloquent on the benefits to be had by using the simulation method, it fails to mention any drawbacks. I can think of one: that this is probably not a low-cost procedure. Is this correct, and are there other shortcomings?

2. Just how microscopic does one want to be? Is there no danger of being flooded with numerical output? Might it not happen that simulation of sufficiently complicated relationships would take more time than the corresponding events take in real life? How serious would this be?

3. How do you generalize your numerical results to get meaningful general propositions?

4. It would be most helpful if the problems and prospects of simulation could be related more closely to the subject of regional studies and regional accounting. Could Mr. Orcutt give some examples, perhaps from his forthcoming book,[5] in which simulation has been applied to problems with a regional flavor, or could he discuss some regional problems for which simulation appears particularly suitable?

If these issues cannot be settled at this Conference, I am sure that Mr. Orcutt's book will have a lot of interesting things to say about all these matters.

## *J. W. Petersen*

*The RAND Corporation*

Mr. Orcutt has stressed the advantages of microanalytic models for economic study. I am in general agreement concerning these advantages; however, I think it is important to be aware of some

[5] Orcutt, et al., *op. cit.*

of the basic problems that arise as one attempts to use such a model.

During the past several years I have had an opportunity to participate in the design and operation of three major microanalytic simulation models. They are difficult models to use and, like all powerful tools, they are easy to misuse. Let me touch on several of the problems involved.

It is not true, as Mr. Orcutt indicated, that in a microanalytic model as many variables as desired can be used. First, we usually do not know enough to specify relevant relationships with any real confidence; second, the introduction of additional variables increases research costs at a frightening rate. This latter point almost always limits our endeavors. As a result, any practical microanalytic model will be a compromise of the pure model outlined by Mr. Orcutt, and to take on any significant problem will require the aggregation of many microrelationships. The simplifications implied in such aggregations produce a simulation which is, as it always must be in economics, a severe abstraction of reality. This means, in turn, that all participants in the project must be aware of the relevant abstractions; and hence they do not, as Mr. Orcutt suggests, have "nearly complete freedom in specifying the operating characteristics of decision units and other components." It is easy to become mesmerized by any model, but particularly with a large simulation model. Project participants must remain aware of model short cuts and limitations, since, without such awareness, interpretation of model results can be spurious. Because a simulation project is generally large, it becomes imperative that the model development and interpretation be very closely co-ordinated. This would seem to rule out any naive "building-block approach."

The key problem of a large simulation model lies in the difficulties of translating simulation results back into reality. The model is an abstraction, and because of the complex relations that can be handled the researcher is hard pressed to separate simulation artifacts from meaningful results. In our current project, which is a simulation of Air Force support operations, we have attempted to overcome this problem by putting knowledgeable men into the model at key decision points. By constraining them to operate under a prescribed set of rules it is possible to maintain the essentials of an analytic model as contrasted to gaming or a man-machine study. The only reason we have men in our model is to obtain their comments concerning the reality of our simulation. By this means we have an opportunity for systematically improving the model and, at the same time, have a basis for viewing model results with greater confidence.

Let me conclude by making several statements which I will not attempt to justify except to say they are based on experience with microanalytic models to date.

1. I don't believe simulation models are particularly useful in problems of basic policy determination. All such questions boil down to trading off the costs and benefits of alternative actions and such questions are handled more cheaply, more easily, and better by the more traditional tools of aggregate analysis. Micro models are too big and expensive to use in a Monte Carlo sense—awkward at best.

2. Simulation techniques are best used for problems of policy implementation. Given basic goals or policies, simulation techniques have proven very useful in developing and preservice testing operating control systems.

3. The development of an operational simulation model is quite an experience. The method is a severe taskmaster—it just doesn't allow you to sweep as many knotty problems under the rug. That is, by its very nature it makes you dig deeper, and in so doing allows you—almost forces you—to better insights into the problem you are studying. This is its real advantage!

Someone once said with regard to the simulation technique—you get all the advantage you're going to get just by putting the model together; don't bother running it. With the exception of developing control systems, I think this is true.

9

# THE ROLE OF ACCOUNTS IN THE ECONOMIC STUDY OF THE PITTSBURGH METROPOLITAN REGION

*Edgar M. Hoover and Benjamin Chinitz*
*University of Pittsburgh*

THIS PAPER aims to help clarify the part that systematic regional accounts can appropriately play in basic economic studies of metropolitan regions. Our views have been shaped largely by experience in the design and early phases of execution of the Pittsburgh Regional Planning Association's three-year study of the Pittsburgh region, and in the latter part of this paper we shall refer specifically to that design. We hope, however, that this methodological discussion will be of wider interest and applicability.

## *I. GENERAL FUNCTIONS OF ACCOUNTS IN THE ECONOMIC ANALYSIS OF A REGION*

It is well at the start to set up our notion of what comes under the term "accounts" in the context of this paper. Accounts mean, to us, systematic and quantitative cross-tabulations of economic transactions and claims, primarily in money units and primarily cross-sectional in nature. Basic to this notion of accounts is the feature of double-entry: every transaction and every claim involves two parties, and in principle each item in a set of accounts is cross-checkable because it occurs twice. The "parties" in a set of regional accounts are not individual households or firms, but appropriate groupings thereof, which we shall call "sectors." Finally we propose to confine our attention in this paper to transactions (debits/credits) accounts only, ignoring the whole area of "balance-sheet" or asset/liability accounting.

Accounts, as just defined, are useful at two different stages of regional analysis. Though the distinction is perhaps not so clear-cut as we here (for simplicity's sake) make it appear, we can separate two major roles of accounts under the heads of "consistency-checking" and "impact evaluation."

### *Consistency-Checking*

It is clearly important that any analysis avoid internal contradictions so far as possible. Double-entry bookkeeping is an eminently logical way in which to impose that constraint upon our educated guesswork.

Three types of consistency-checking suggest themselves, though perhaps only one of these really involves accounting in a legitimate full-fledged sense:

a) Identity of the two sides of a single intersector transaction flow; i.e., *A*'s payments to *B* are the same as *B*'s receipts from *A*. Symbolically we might write this as:

$$\underset{A\quad B}{Cr} = \underset{B\quad A,}{Dr}$$

or

$$\underset{A\quad B}{Dr} = \underset{B\quad A.}{Cr}$$

*Example:* expenditures by residents at regional drug stores are identical with sales of regional drug stores to residents. This is of course tautological; such identities may be useful reminders but in themselves are merely building blocks for full-fledged accounts.

b) Summation of one side of a sector's accounts; i.e., *A*'s total payments are the sum of *A*'s payments to all specific sectors. Symbolically,

$$\Sigma_A Cr = {}_ACr_B + {}_ACr_C \ldots . + {}_ACr_N.$$

*Example:* a synthesis or breakdown of the region's personal income. This is also tautological if the total is arrived at simply by measuring all the components and adding them up. The summation-type equation is really useful only if there is some independent way of estimating the total, and if some components (but not too many) are impossible to measure directly with accuracy. In this case, the hazy component(s) can be derived as a residual—or rough independent estimates of its size can be checked for consistency.

c) Balancing the books of a specific sector. Symbolically,

$$\Sigma_A Cr = \Sigma_A Dr,$$

or [cf. (a) above]:

$$\Sigma Cr_A = \Sigma Dr_A.$$

Here we have a full-fledged sector-accounting constraint, capable of making a major consistency contribution and of being built into an over-all system of accounts describing the intersector transactions of the economy.

*Example:* the regional balance of payments.

It has already been suggested that a real consistency check requires some direct estimation of all components in an aggregate—unless a residual is moderately small and subject to some appraisal of its plausibility, the account of which it forms a part cannot be used to impose consistency in any real sense.

The essential point here is that as we depart substantially from exhaustive accounts, the usefulness of the accounting procedure very rapidly dissipates, so far as consistency-checking is concerned. This sharpens the question a regional economic study design has to face. The effort devoted to filling out accounts must be of at least critical minimum size, or reliance on accounts *per se* to impose a consistency check must be foregone.

### *Impact Evaluation*

This includes the identification and measurement of relationships among the various sectors of the regional economy, in such fashion that the results of any expected or supposed change upon the economy can be traced through the series of direct and indirect repercussions.

We have two points to make here regarding the potential usefulness of comprehensive regional economic accounts. First, their usefulness can be minimal if they are done on a one-shot basis. Second, even full and current accounts would not provide an adequate apparatus to generate the results a regional economic study should provide. Let us examine these two points separately.

*Need for Continuous Accounting.* Before a regional analysis can be of substantial real use to planners and policy-makers it must take on some dynamic character. A one-time cross-sectional description of the "economic base" or of the trade flows and other connections among sectors of the regional economy is only a first step (albeit an essential one) *toward* a useful analytical apparatus.

It is clear in this context that if accounts are to play an effective part, they must be kept more or less current, and must cumulate as time series. The national income accounts would lose most of their manifold utility if they were not compiled as continuing time series. Regional accounts, too, will have to be put on a continuing basis if they are to serve as the basis for evaluation of changes over time. Most of the payoff comes not when the first balancing of books is achieved, but at a much later stage.

The preparation of a comprehensive regional accounts system is therefore an appropriate task for a continuing statistical agency, and not for an *ad hoc* project with limited life span.

*Adequacy of Accounts for Tracing Impacts.* Perhaps the most frequently suggested operational *application* of regional accounts (as distinct from their direct consistency-checking role) is in the evaluation of "multiplier" effects. The question here addressed is, "What changes in total regional business activity and income will result from a specific change in the region's earnings from export of goods and services?" Coefficients derived from interindustry and intersector transactions accounts are used in tracing repercussions of the initial change. The additional receipts from exports are distributed to businesses (for materials, supplies, and services), to government (as taxes), and to households (for labor services or as proprietary income). Out of such parts of these payments as are received within the region, further payments for goods and services are made by the recipients; and so on. At each round, a part of the new flow of payments leaks out of the region, and the ultimate total increment to regional activity and income is derived as a function of those leakages.

The attempt to derive a useful multiplier along such lines is, of course, beset by special difficulties when we deal with an open regional economy. The interindustry and intersector coefficients are not only difficult to determine, but may well be even less stable than the corresponding relations for the national economy. The most serious difficulty, however, is with the "leakages." The input side of each technical coefficient has to be broken down into inputs from internal versus outside sources, and that breakdown, even if once determined on a cross-sectional basis, is certainly in most instances highly unstable.

One may question how useful such a multiplier calculation would be even if it were based on precise and complete accounts of all the transactions involved. It is important to note that only a quite limited range of repercussion effects (in terms of locational analysis, only those linkages involving "market orientation") is actually subject to evaluation in this way. Our impression is that the importance of the multiplier effect for real policy and planning guidance has been oversold. In practice, only very occasionally does a major decision in private or public planning rest on a question like, "If we get (or lose) this export establishment, by how much will total local employment and payrolls ultimately be raised (or lowered)?" Instead, we are faced with questions like the following:

1) What industrial and population size and structure should we assume five, ten, or twenty years ahead as a basis for investment or other decisions that have to be made now?

2) Where in the region will changes occur? (This geographical aspect is almost always crucial.)

3) What effects on growth or internal patterns are foreseeable if (for example) we:

a) put through a renewal project,
b) change the tax structure,
c) improve the skills of the labor supply,
d) redistribute the costs of internal passenger movement,
e) impose stricter pollution controls,
f) negotiate lower freight rates,
g) attract the headquarters of a big national corporation,
h) build convention facilities, or
i) increase the stature of our universities?

The "multiplier" effect and transactions-accounting play a quite minor role in an analysis adequately responsive to such questions as the above regarding repercussions of an assumed exogenous event. They play no role at all in the explanation or prediction of the exogenous event itself.

### *Alternative Emphases in Regional Economic Analyses*

The discussion above has shown (1) that development and maintenance of a comprehensive system of accounts for a regional economy is an even more formidable task than might at first appear, if the system is to be sufficiently complete, current, and historically grounded to function adequately; (2) that regional economic studies have to provide answers on a great many questions in which accounts could at best play a minor role; and (3) that the time and resources available to such studies and their responsibilities, as distinct from those of continuing statistical agencies, limit the extent to which they can appropriately commit effort to development of accounts as such.

All these considerations bear on the way in which the research effort of such a study can best be focused. Later in this paper we shall set forth the approach charted for the Pittsburgh Study. But first it will be helpful here to consider a more general dichotomy of approaches. The two approaches we shall characterize in terms of "vertical" vs. "horizontal" emphasis.

By "vertical" we mean an approach that seeks to probe as deep as possible into the workings of the economy of the region under study, while the outside world as such is not studied or even observed directly. Only transactions with the outside world are recognized, as items in the accounts of the region in question.

What we call here the "horizontal" approach leans heavily on (1) interregionally comparative structural analysis, and (2) locational analysis in terms of interregional competition and complementarity.

The vertical approach addresses itself to a much narrower range of questions than the horizontal, since it takes all events or influences emanating from the outside world as exogenous, and merely seeks to provide a mechanism for evaluating the internal impacts of certain types of events that may be assumed to occur, such as an increase in exports. The "multiplier" type of question already cited is illustrative of the range of inquiry appropriate to this approach. The horizontal approach, by contrast, pays a great deal of attention to the "outside world," and seeks to exploit another dimension of the mass of available data: namely, data for other regions either (1) similar to, or (2) economically tied to, the region under study.

Groups of *similar* areas are studied in order to discover empirically some useful trends, regressions, and other statistical relationships that form parts of the explanation of past or prospective behavior of the region we are studying. For example, in seeking to understand the levels and patterns of consumer and business service activities in the Pittsburgh Region, we have found it helpful to examine statistically the ways in which such activities are associated with total employment, population, income, retail sales, age structure, and industrial specialization, etc., among the medium sized and larger metropolitan areas of the United States in years for which data are readily available. This is quite different from the "vertical" approach, which would measure the receipts and outpayments of each of these types of service establishments in the Pittsburgh area and try to develop from such accounts some rules about how levels of local service activities would react to, say, changes in total Pittsburgh income, or Pittsburgh exports.

Secondly, the "horizontal" approach considers specific regions other than, but *economically related to,* the region under study. This is in connection with interregional locational analysis, which is aimed at explaining and projecting the region's activity levels in what are often called "basic" industries. All this is exogenous to the "vertical" approach *per se*—accounts for a region are not generally designed to *explain* the levels of the basic activities. The outside regions to which analysis is extended in the "horizontal" approach include primarily regions related to ours either as markets or as competitors. For example, the Pittsburgh level of output of one of our specialties, oil-country goods, depends largely on (1) the growth of demand of certain oil and gas producing areas, particularly in the Gulf region, and (2) changes in Pittsburgh's cost position relative to two or three areas that offer its principal competition in serving those markets.

As suggested earlier, the development and use of comprehensive regional accounts go with the "vertical" approach. Accounts are de-emphasized in the horizontal approach. Partly this is because many regions, rather than one, are involved, and no one has yet seriously suggested that it is feasible to carry out an operationally adequate metropolitan economic study in the framework of a multi-region interindustry model of the Leontief, Isard, Moses or other types. But also, accounts play a minor role in the horizontal approach partly because the use of that approach implies, as a rule, a strong interest in evaluating economic links and relationships on which accounts themselves throw little or no light, and which are outside the scope of a vertical-type investigation. Also, the horizontal approach dispenses with a great deal of accounts effort because it provides acceptable (and often easier) ways of accomplishing some of the same purposes.

This last point brings us back to the fact that the two approaches (one in effect maximizing the use of accounts and one minimizing it) are, to an important degree, alternatives. We have heard of no instance in which a metropolitan economic study had the resources to push both approaches simultaneously to the limits of their productiveness. The typical if not inevitable situation is one calling for a major choice between "vertical" or "horizontal" emphasis in the allocation of relatively limited resources.

The choice will depend on many factors, not excluding the analytical predilections of the study designers. One important consideration is the necessary commitment of resources for those study objectives (e.g., analysis of the interregional location factors affecting the levels of basic industries in the region) which must in any event use a "horizontal" approach. If enough resources then remain to develop also the vertical approach through an adequate set of regional accounts, and if there is a prospect that such accounts can be kept current, that effort may be advisable—though the possibilities for either supporting or shortcutting the accounts task by horizontal analysis of comparative economic structures in similar areas may still be considerable.

## *II. ROLE OF ACCOUNTS IN THE PITTSBURGH STUDY DESIGN*

In the foregoing section we have outlined some of the criteria on which it seems appropriate to decide how much effort should go into accounts work, as such, in a regional economic study. Now we can apply these criteria in the specific context of the Pittsburgh Study design.

The Pittsburgh Study was set up with a $400,000 budget for a three-year period ending in 1962. Its objectives have been broadly laid down as follows:

1) To analyze the economy of the Pittsburgh metropolitan region (comprising six counties in southwestern Pennsylvania) and of areas within it; to identify the principal economic factors in the region's development; to provide a basis for anticipation of economic change in the Region and for sound private and public planning.

2) To lay the groundwork for a continuing process and organization which regularly thereafter will take the pulse of the regional economy, and which may serve as a pioneer model for similar undertakings in other metropolitan regions.

3) To advance the scientific study of regional economics.

Analysis of subareas within the region is an important part of the study. We propose to ignore that in this discussion, however, and to focus on the role given to accounts in analyzing the economy of the total area. It may be remarked in passing that any systematic effort to deal with the internal spatial pattern in terms of accounts seems wholly unfeasible.

The Study is expected to provide some long-term projections of growth and change in the region in terms of population and its characteristics, labor force, employment by types, and various measures of well-being such as level, distribution, and stability of income, levels of public services and housing. It is designed to dovetail with independent specific studies on internal transport, urban renewal, the central business district, and other topics.

The Study is also expected to provide a proposal for following up the basic three-year study with a continuous system of economic data collection, analysis and reporting. Our discussion in this paper does not refer to that eventual system, in which it may be presumed, for reasons already adduced, that accounts will play a relatively larger role than in the three-year basic study, and an increasingly important role as time goes on.

In terms of the general discussion of approaches in Part I of this paper, it is clear that the Pittsburgh Study involves a large amount of work on concepts and relationships that fall outside the scope of transactions accounting and outside of what can be attacked in a "vertical" approach. The Pittsburgh Region is among the most specialized of the large U.S. metropolitan areas.[1] In explaining the Region's growth and structure and projecting changes in it, the more critical and larger part of the task runs in terms of the region's competitive position, *vis-à-vis* rival regions, in certain basic industries and in the implications of the existing Pittsburgh "mix"

[1] Among at least the fifteen largest Standard Metropolitan Areas, only Detroit has a greater degree of "industrial specialization" than Pittsburgh, as measured by the share of the area's total manufacturing employment accounted for by the largest single 2-digit industry. Donald J. Bogue and Evelyn M. Kitagawa, *Suburbanization of Manufacturing Activity within Standard Metropolitan Areas*, University of Chicago, Studies in Population Distribution, No. 9 (Scripps Foundation, 1955), pp. 35-38. Based on data from U. S. Census of Population, 1950.

of activities that are, nationally, growing at faster or slower rates. Detail and attempted precision in the tracing of impacts of these trends upon local trade and service activities are of decidedly secondary importance.

This in itself suggests that such a study could not appropriately devote the bulk of its resources to the development of a comprehensive set of regional accounts for the sake of the light such a system might shed on internal repercussions of exogenous change.

As to the contribution of accounts to internal consistency, our previous discussion suggested that no real consistency check is forthcoming unless the accounts are built up from substantially complete sets of measures of components of the transactions flow. A "critical minimum" scale of effort is involved.

The time and money cost of this critical minimum effort is, we believe, large in the context of presently available data for regional accounts: so large that we decided that we could not attain it and still have sufficient resources for the nonaccounts work needed in order to derive useful findings of the order which we are committed to supply.

This decision may have been correct or incorrect. A contrary decision—to go all out for an adequately tight accounting—may well be justified in other studies, particularly where (a) the budget is larger relative to the size of the area, (b) more extensive accounts data, or collection channels therefor, are already at hand, or (c) the objectives and mandates of the study are less concerned with early policy-guiding payoff and more with laying statistical groundwork for subsequent study over a longer period. It is also fairly certain that, as time goes on and data are improved and accumulated, attainment of the critical minimum degree of regional accounting coverage will become more economically feasible.

In the three-year basic study, then, we are not attempting development of regional accounts in a thorough or comprehensive way. We are relying largely on alternative methods for quantifying relationships and checking internal consistency, and developing accounts on a partial and selective basis. To appreciate the role they do play in our design we must indicate, more specifically than has so far been done, the main lines of our analytical scheme.

We start by grouping the multifarious activities of the regional economy into meaningful sectors. The purpose is to distinguish important measurable and relatively homogeneous groups between which there are notable differences in respect to:

1) activity-level determinants;
2) nature of impact on other group activity levels or economic welfare indices (income, stability, amenity) in the region.

The types of activity we distinguish are outlined below:

*Endogenously Located Activities*—i.e., activities adequately explainable on the basis of other activities, groups of activities, or other economic indicia within the Region itself.

a) Consumer trade and service activities (characterized by close market orientation to households, so that information regarding households (number, income, other characteristics) is the basis of explanation and projection of the activity level for the category. Some activities in this group show consistent concentration (relative to population and income) in larger metropolitan areas, because of scale economies, differences in consumer demand patterns, etc. We can still categorize them, though, as explainable in terms of Regional household data.
b) General business and professional services, closely market oriented to a wide range of activities centering in the Region.
c) Government activities, for the most part, will behave like either (a) or (b) and can be analyzed on a parallel basis; but will be kept separate from nongovernment in each category.
d) Business activities closely linked (in terms of Regional activity levels) with some specific other activity. These can be *grouped,* to the extent that they are linked to a single activity or to specific activities which are themselves grouped together in our classification scheme. The linkage may be any of various types; it is not confined to close contact or market orientation.
e) Households—on the hypothesis that population growth can be explained and projected on the basis of employment opportunities, rather than the other way round.
f) Construction.

*Exogenously Located Activities*—i.e., activities *not* adequately explainable in terms of other activity levels or economic indicia for the Region alone.

a) Some types of state and federal government outlay.

b) Business activities in the production of goods and services in which substantial interregional trade exists (i.e., there are substantial interregional cost differences, and the product is interregionally transportable at low enough cost to allow these cost differences to produce substantial trade and specialization among regions). This category of activities corresponds roughly to what are generally called the "basic" industries—but we include in it those in which the Pittsburgh Region is a substantial *importer* as well as those in which it is a substantial exporter.

Within this "exogenously located business activities" category, we make a further disaggregation along these lines:

i) A number of very important industry complexes for which we are making special intensive studies. These lie in the metals and metals products, stone-clay-glass, and chemicals fields. For each of these we expect to develop a maximum of information on cost and market factors affecting location; on local inputs and outputs; on transactions with the outside world; and in impacts (long-run, cyclical, and seasonal) on the local economy, both directly and in terms of local transactions and indirectly in terms of effects on the regional labor and land markets, etc.

ii) One or more groups of activities for which Regional activity levels can be adequately explained primarily on the basis of labor supply.

iii) One or more groups of activities for which Regional activity levels can be adequately explained primarily on the basis of the Region's importance as a central-office focus.

iv) Such other groupings as it may be found appropriate to set up, on the basis of a common dominant exogenous factor determining the Region's activity level.

Of the two sectors, endogenous and exogenous, the former is of greater interest for the present discussion because of the role which accounts typically play in the analysis of that sector. We shall therefore not elaborate here our methodology for the exogenous sector. Suffice it to say that we expect the exogenous-sector analysis to absorb a very substantial part of the Study's resources and to contribute in corresponding measure to achievement of our ultimate goals. When we have projected all the exogenous components (including most of manufacturing, mining, research, and central-office activity) we shall have covered 40 to 50 per cent of the total economy without recourse to any elaborate simultaneous-equation approach.[2]

How then do we pass from a projection of exogenous sectors to a projection of endogenous sectors, and hence of the whole economy?

We have three primary projection targets: population, employment, and personal income. All other magnitudes can be thought of strictly as "intermediate products." If they can be bypassed in the analytical framework, nothing essential is lost. In each case—population, employment, personal income—we shall have national projections and also projections for parts of the nation (e.g., states and all SMA's) developed by outside sources, one of which, the National Planning Association, is being supported in part by Study funds to that purpose.

First, as to projections of population: In our model, as already noted, we assume that *population is a function of employment.*[3] But what is the function? From national projections of labor force participation by age and sex, we can derive general indications of trends in the disposition to seek work. But we have learned that Pittsburgh has special features which have made it depart from the national "norm" in the past. One of these features, apparently,

[2] This should not be taken to mean that we pay no attention to the impact of Regional steel-industry wages on apparel-industry wages, the impact of steel-industry employment on Regional labor supply, etc. But these types of interdependence cannot generally be handled in a mathematical model framework in any event.

[3] Subject to possible constraint or qualification on the basis of our analysis of trends in migration and other demographic factors.

is the industry mix; others will be identified. Demographic projections will indicate the effect on over-all Regional labor force participation likely to be executed by changes in the age and sex composition of the population. Provisionally at least, then, we may think of a functional relationship of this type:

$$P = f\ (D, E_M, L_N)$$

wherein:

$P$ = population in the Pittsburgh Region,
$D$ = demographic characteristics of the population in the Pittsburgh Region,
$E_M$ = the industry mix in the Pittsburgh Region,
$L_N$ = the national labor force participation rate.

Now let us consider a typical endogenous employment sector, say, wholesale employment. The dependent variable is *per capita employment in wholesaling.* Again we begin with the national projection. This provides us with an indication of how rapidly the function is expected to grow. It embodies all the effects of changes in demand, productivity, etc., in one series. We know here, too, that "Pittsburgh is a bit peculiar" and expect in due course to have identified the critical variables "explaining" these individual characteristics of the Region. Since wholesaling serves both population and industry, one could visualize a regression equation of this type:

$$W = aM + bR + c,$$

wherein:

$W$ = wholesale employment per capita,
$M$ = manufacturing employment per capita,
$R$ = retail employment per capita.

Referring back to our earlier discussion, it is apparent we expect to develop such functional relationships by the "horizontal" approach. National projections of changes in both dependent and independent variables will be used as one guide to changes in the parameters of the functional relations.

The projection of retail employment per capita brings personal income into the reckoning. Here again we shall have national and subnational (e.g., state and aggregate metropolitan) projections to start from on the growth of retail activity per capita, and also past and current cross-sectional analysis throwing light on the factors influencing per capita retail employment in the Pittsburgh Region relative to comparable areas. Again, we have to take account of the Pittsburgh Region's departure from "norm" in the past. Part of the explanation will run in terms of levels of personal income. Other parts may involve income distribution, household size, etc. Some of the factors introduced to explain Pittsburgh's individual retail behavior (e.g., distance to nearest larger metropolis) might remain unchanged in the projection period. Those that are subject to change must be either solved within the framework of the model or determined exogenously. Personal income is inside the model. Ideally, distribution of income should also be. But many quirks of consumer behavior will probably have to be handled outside the model.

What we are offering is, in effect, a more articulated and refined version of the basic-nonbasic approach. That is to say, while the basic-nonbasic approach attempts to relate total to endogenous employment—necessarily a highly unstable relationship—we work with a whole series of such relationships. We improve upon the crude ratio approach by disaggregating not only the variable on the left-hand side of the equation, but also the variable on the right-hand side. At the same time, we stop short of disaggregating to the degree that is implied in most input-output tables.

Some sectors—how many, we do not yet know—will not be susceptible to this type of treatment. For example, construction employment may not be adequately explainable as a function of total employment, and/or changes therein, even when total employment is somewhat disaggregated. It may be necessary here to come to grips with the sources of demand, i.e., residential construction and business capital formation. Government employment is another case calling for special treatment.

It should be obvious that in this kind of framework there is no such exhaustive cross-tabulation of flows as is found in typical regional accounts systems. Flows are identified only as a last resort.

Only fragments of accounting tables, in general, emerge. Nevertheless, there are a few places in the model where the build-up of substantial aggregations of data in accounts form is implied.

One of these is personal income, which would be a crucial magnitude if only because of its significance as a welfare measure. We could try to bypass the whole question of how personal income is derived by making its over-all level a function of employment, possibly disaggregated to take account of the differential income yields of different industries. But even if a short cut like that would work for the purposes of the model, there would still be much intrinsic interest in the *sources* of personal income: how much in wages and salaries, how much in dividends, what is the net transfer payment position of the Region, etc. Hence, we do expect to build up a fairly systematic accounting of personal income by source.

Another example is the government sector. Federal, state, and local payments in the Region need to be distinguished, with breakdowns of expenditures by major services and transfer categories, as part of the measurement of welfare.

We do not expect, however, to emerge with complete accounts of any of the following more or less standard types:

gross regional product,
the regional balance of payments,
the regional investment and savings account,
income produced within the Region,
income received within the Region,
the interindustry transactions matrix.

None of the six accounts just listed is of direct use *per se* in making evaluations and private or public planning decisions of the sort the Study is designed to underpin. All are highly difficult to build up as exhaustive and consistent aggregates, because of some substantial components (especially those involving financial transactions across the borders of the Region) which defy accurate measurement. The complete input-output matrix presents the further difficulty of enormous detail (especially if each input and output is split between Region and outside world).

Consequently, we do not believe it feasible, in the context of our Study, to "account" these items. Such large portions of each aggregate are so difficult to measure directly that any consistency-checking value of an attempted account reconciliation becomes illusory. We are, of course, free to derive estimates of any of these aggregates we want, and any of their components, by eclectic "side calculations," but there will be large residuals whose reasonableness we can hardly expect to appraise.

## COMMENT

*Britton Harris*

*Penn-Jersey Transportation Study*

In discussing the Hoover-Chinitz paper, I shall find it necessary to refer back to certain very elementary concepts of regional interaction.

Let me first say what I am *not* going to debate. I am not going to discuss the distinction which has frequently arisen between conceptual and operational models. In my mind, this distinction is a thin one because a conceptual model must be tested to be validated and at this time it becomes operational. I am also not going to differentiate between "pure" research and research which is policy oriented. To my mind both of these types of research attempt to answer predictive questions—that is, to specify the consequences of changes in conditions. Frequently, however, as in the case of the Borts-Stein model, we find predictions based on proximate causes whose analysis leads us directly into the basic theory of regional interaction and change.

Broadly speaking, these sessions have been based on an implicitly agreed-upon theory of regional interaction which consists of three parts. First, we consider that each region has certain distinctive advantages over other regions in carrying on certain activities. Second, we regard these advantages as offset by the costs of overcoming distance. Finally, we regard these two counteracting forces as giving rise to an equilibrium interchange of goods and services based upon the structural features of the economies of all regions.

When we view the outlook of the Pittsburgh Study in the light of this theoretical scheme, we readily observe that the future comparative advantage of the Pittsburgh Region is by far the most important element in a long-term projection. The construction of a local input-output scheme would only help to specify more precisely the consequences which would flow from the projected changes in regional advantages and in the location of "basic" activities. From this it follows that no apology is required on the part of the Pittsburgh Study for a preoccupation with the analysis of regional advantage and its projection into the future.

It is clear from this example, so well developed by Hoover and Chinitz, that regional accounts should serve the purpose of estimating regional advantages and trends therein, and verify theories as to the consequences of these advantages by actual measurement of flows in interregional trade. Thus the accounts should display resources, factor costs, and movement volumes. Factor costs in this array would necessarily include transportation costs.

In these discussions we have paid relatively minor attention to the problems of the measurement of flows of goods and services. These flows (as has been implied by our avoidance of them) are difficult to measure. It is to be hoped that the basis for an adequate transportation model, connecting interregional flows and regional activity and making it possible to deduce the former from the latter, may in the future be laid by means of a much needed national census of transportation which will provide at the same time the necessary information regarding transportation costs.

One might argue on principle against the necessity for a study like the Pittsburgh Regional Economic Study to review national location patterns, factor costs, and goods movements—on the grounds that such effort is duplicated from one study to the next. This is indeed true, but if the general framework of regional accounts does not provide this information such duplication will be forced upon various successive studies. We are, therefore, discussing accounts in the framework of which general regional analysis can proceed, and this framework must include adequate information on transportation and the flow of goods and services.

In this general context, I have very few particular comments to make beyond commending the authors of this paper for their realism and insight.

I find some fault with the frequent references to the instability of trade relationships as between regions, and even more with the reference to the instability of the relationship between total activity and endogenous activity. In the former case, relationships tend to be very stable in most cases over long periods of time, and to focus correctly on the sudden changes which may take place is not to deny their stability. In the latter case, the relationship (sometimes unfortunately called the "basic service ratio") is in fact stable, and more stable than the trading relationships. In fact the Pittsburgh Study depends on this stability as justification for the lack of attention which it gives to local input-output relationships.

## *James M. Henderson*

*Upper Midwest Research and Development Council, University of Minnesota*

The Hoover-Chinitz paper is separable into two distinct parts. The first contains a discussion of why regional accounts are not very useful for the Pittsburgh Study. Unfortunately, this discussion is cast in negative terms. The authors will not construct detailed input-output or income accounts, and no one could insist that they should, given their resources and requirements. Their research, however, will contribute to regional accounting. The authors mention detailed personal-income accounts. The use of an exogenous-endogenous approach will require more flow accounting than they suggest. Flow information will be necessary to tie the outputs of Pittsburgh's export industries to the future demands in their major markets. Furthermore, flow information may be necessary to determine which sectors are exogenous. The real choice is whether flow information should be collected and organized within a systematic framework or whether it should be developed in an uncoordinated manner in the investigations of particular industries.

The second and major part of the Hoover-Chinitz paper is devoted to a partial description of their "horizontal" approach. The discussion here is limited to several, hopefully constructive comments.

The use of cross-section data for individual metropolitan areas to estimate parameters relevant for Pittsburgh presents some new statistical problems. One could imagine a situation in which a regression analysis yielded parameters with small standard errors, but in which the Pittsburgh observation made a relatively large contribution to total variance. In addition to using the conventional measures of statistical reliability, Hoover and Chinitz must develop methods to determine whether their regressions provide adequate "explanations" of the Pittsburgh observations.

The two equations presented for illustration provide an adequately vivid indication of the simultaneous nature of the relationships with which Hoover and Chinitz will deal. Population is a function of disaggregated employment. The ratios of employment in some sectors to population are, in turn, functions of the ratios of employment in other sectors to population. The authors face a dilemma. Their system contains simultaneous relationships. They will use multiple regression for some equations, but an elaborate simultaneous-equation approach does not appear feasible. Some more mundane approach, however, might do the job.

If the authors can adequately describe the Pittsburgh economy in terms of a sequential set of regression and definition equations which can be solved one at a time, their dilemma may be resolved. The following provides an example of a sequential set:

$$\begin{aligned}
X_1 &= \alpha_{10} && && && + \alpha_{51} Y_1 + \alpha_{61} Y_2 + \epsilon_1 \\
X_2 &= \alpha_{20} + \alpha_{21} && X_1 && && + \epsilon_2 \\
X_3 &= \alpha_{30} + \alpha_{31} && X_1 + \alpha_{32} && X_2 && + \alpha_{53} Y_1 + \epsilon_3 \\
X_4 &\equiv && X_1 + && X_2 + X_3 &&
\end{aligned}$$

Where $X_j$ are endogenous variables, $Y_j$ are exogenous variables, $\alpha_{ij}$ are regression parameters, and $\epsilon_i$ are error terms. The first three are regression equations and the fourth is a definition. Given values for the exogenous variables, the first equation yields an estimate for $X_1$. Given the estimate for $X_1$, the second yields an estimate for $X_2$. The third then, in turn, yields an estimate for $X_3$, and so on. Additional effort is necessary to determine the statistical properties of the above schema since it violates some of the classical multiple-regression assumptions.

The Pittsburgh Study presents many challenges, but it also offers promise of a substantial breakthrough in regional analysis.

# AUTHOR INDEX

# SUBJECT INDEX